AVOIDING ARCHAEOLOGICAL DISASTERS

AVOIDING ARCHAEOLOGICAL DISASTERS

A RISK MANAGEMENT APPROACH

DARBY C. STAPP AND JULIA G. LONGENECKER
with contributions by
Roderick Sprague, Thomas F. King,
Michael S. Burney, Mary Rossi, and Adrian Praetzellis
Illustrations by Anthony Smith

Routledge
Taylor & Francis Group
LONDON AND NEW YORK

First published 2009 by Left Coast Press, Inc.

Published 2016 by Routledge
2 Park Square, Milton Park, Abingdon, Oxon OX14 4RN
711 Third Avenue, New York, NY 10017, USA

Routledge is an imprint of the Taylor & Francis Group, an informa business

Library of Congress Cataloging-in-Publication Data is available.
Stapp, Darby C., 1954–

Avoiding archaeological disasters : a risk management approach / Darby C. Stapp and Julia G. Longenecker ; with contributions by Roderick Sprague ... [et al.].
p. cm. — (Techniques and issues in cultural resources management)
Includes bibliographical references and index.
ISBN 978-1-59874-160-5 (hardback : alk. paper) —
ISBN 978-1-59874-161-2 (pbk. : alk. paper)
1. Cultural property—Protection. 2. Historic sites—Conservation and restoration. 3. Construction industry—Risk management. 4. Project management. I. Longenecker, Julia G. II. Title.
CC135.S73 2009
363.6'9—dc22

200902884

ISBN 978-1-59874-160-5 (hbk)
ISBN 978-1-59874-161-2 (pbk)

Cover design by Andrew Brozyna

Contents

Illustrations

Tables

Preface

"What was that?" the track hoe operator yelled as he pulled back the bucket. "I don't know, probably rocks. Keep going, we've got to get this done today," snapped the supervisor. Digging continued, and minutes later, the back dirt pile was littered with wood and rocks, broken pottery, and bones. "You better stop work, Joe; I have to call this in."

We hear stories like this one time and time again. A construction project stops because of an archaeological discovery. Time stands still while the project proponents, archaeologists, bureaucrats, and concerned groups hold meeting after meeting to decide what to do next. Costs escalate, schedules evaporate, and people become frustrated, angry, and confused.

Archaeological discoveries during construction projects occur fairly frequently. Some unknown ancient Roman temple is discovered during construction in Naples, Italy, or a prehistoric cemetery is discovered during road construction in Phoenix, Arizona. Projects around the world routinely run into archaeological issues for one very good reason. Places that people choose to use today are often the same places people used in the past. Certain places are good places to live, and people keep coming back to them, from prehistoric times to the present. If your project is located in one of these places, odds are high that something will be discovered at some point during the life of the project.

Most archaeological situations are manageable; either nothing important is found within the project area, or if something is unexpectedly discovered during the planning phase or even during construction, the parties involved are able to figure out how to manage the situation and get the project back on track. There are times, however, when these situations are mismanaged, and this breakdown can lead to chaos. Common causes of mismanagement include:

- project managers ignore recommendations to conduct archaeological and historical assessments before purchasing a development site;
- project managers fail to contact archaeological oversight agencies to seek their impressions and recommendations;
- project managers contract with archaeological firms that fail to perform;
- and project managers fail to consult with indigenous groups and other interested parties who may have concerns about disturbing the project area.

But project managers are not always to blame. Sometimes archaeologists make bad calls, sometimes archaeological regulators do not give a project enough attention, and sometimes outside parties are not helpful. The focus in this book, however, is on the actions that project personnel themselves can take to avoid an archaeological problem. We have written this book for project managers, project engineers, construction managers, agency personnel, and archaeological consultants involved in large construction projects. Our goal is to provide enough information and enough archaeological perspective to allow the various parties involved in a project to work intelligently with one another.

To help manage an archaeological situation, we present an approach based on risk management principles common to the engineering and construction industries. The approach is straightforward and not especially costly. We explain how to incorporate these archaeological risk management principles into your project and provide the necessary tools to perform the assessments. The approach is flexible enough to be applied to almost any project in any location.

ACKNOWLEDGMENTS

We have seen many projects succeed and fail in the last 30 years. Each has been a learning experience, and we want to thank the many people whom we have had the pleasure to learn from over the years. We especially want to thank those who have participated in the development of this book. This includes six top cultural resource management professionals who prepared material for the book: Garth Baldwin (Bellingham, Washington), Michael Burney (Taos, New Mexico), Thomas King (Alexandria, Virginia), Adrian Praetzellis (Sacramento, California), Mary Rossi (Bellingham, Washington), and Roderick Sprague (Moscow, Idaho). Many others assisted by reviewing the manuscript at various stages. Providing helpful comments on the entire manuscript were several non-archaeological colleagues who work on construction projects: Roger Wright (Pasco, Washington), Jim Frederickson (Richland, Washington), Marv Kinney (Kennewick, Washington), Paul Michalak (Maryland), and Ron Schalla (Kennewick, Washington). Archaeologists who reviewed the manuscript include Michael Burney (Taos, New Mexico), Caryn Berg (Denver, Colorado), and an anonymous reviewer for Left Coast Press. Those who assisted with the effort to apply risk management principles to archaeological settings include Pat Baynes, Andy Hesser, and Mark Triplett (Richland, Washington), and Richard Badalamente (Kennewick, Washington). We also thank Anthony Smith (Lapwai, Idaho) for designing and producing the figures; Kara Pittman (Pasco, Washington) for general comments and assistance in preparing the appendixes; and Thomas Stapp (Richland, Washington) for helping collect information for Appendix D. Finally, we are very appreciative of Left Coast Press for publishing our work. We thank Mitch Allen, publisher; Jennifer Collier, senior editor; Caryn Berg, series editor; production editor Carol Leyba, and the Left Coast Press production crew.

The Disaster of the Superintendent's Path Marking

The San Juan Island National Historical Park, on the coast of Washington state, commemorates the last border conflict between the United States and Great Britain from 1859 to 1872. English Camp, one of four major sites, was originally an American Indian village investigated in 1950 and 1951 by the University of Washington. Much of the site was left as a protected prehistoric midden. When the adjacent site was occupied as a British Royal Marine Camp in 1859, the prehistoric shell midden became a parade ground, with formal flower gardens, pathways, and the post flagpole. It was investigated as a historic site in 1970 and 1971 by the University of Idaho. During the period between the first two seasons, the Park Superintendent instructed the maintenance man to take his backhoe/front-end loader and cover the pathways, with shell taken from the nearby prehistoric midden, to match photographs from the historic era. For years the visitors pocketed, or turned in to the rangers, artifacts and fragments of human bone from the paths. The really frightening conclusion to this sad story is that the Superintendent never did understand what he had done wrong.

Roderick Sprague
Moscow, Idaho, 2008

1 INTRODUCTION

"When I got to the Port Angeles site that morning I walked into what became a mind-boggling progression of events far removed from what I know to be the Section 106 process. Construction had exposed a thick deposit of shell midden and there was archaeological material everywhere. I was pleading with government agents to stop construction and follow the basic process of Section 106 and I pressed the tribal representatives into action for their own sake. No one listened. The construction company wouldn't stop until I could 'show them' the site was intact—or until I found human bones. I did both in short order."

Garth Baldwin, the first archaeologist to identify
the inadvertent discovery at the Port Angeles Graving Dock project

In 2003, archaeological artifacts and bones were discovered during a $19 million construction project in Port Angeles, Washington, just one week after groundbreaking. After a three-month work stoppage while agencies and tribes wrote agreement documents, work restarted, with archaeologists, tribal monitors, and construction staff working side by side. One year later, after the removal of over 260 complete and 700 partial prehistoric Native Americans burials, the local tribe asked that work at the site stop permanently, and the state eventually agreed. Financial costs exceeded $85 million.

The Port Angeles Graving Dock project is not an isolated case. Consider the following:

- A city expanding its wastewater treatment plant knew that its facility was located on an American Indian burial ground. It went through the legally defined approval process yet still wound up accidentally destroying dozens of American Indian burials and severely upsetting the local tribe. Legal settlements cost the city and the contractor several million dollars each, the project was abandoned, and the city eventually agreed to build a new facility in a new location.

- The construction of a new government building in New York City unearthed an eighteenth-century African-American burial ground and subsequently cost millions more in archaeological and project relocation costs. Project personnel did inadequate background research on the site location, incorrectly assumed that all of the area was disturbed, and failed to predict that the local African-American community would take a strong interest in the fate of the burials.
- A large U.S. government agency partnered with a state agency to construct a boat ramp and facility on the Missouri River. The agencies went through the federal archaeological review process; however, the archaeologist failed to heed the warning of local tribal elders and neglected to perform the basic historical research that would have revealed the cultural significance of the proposed project location. When construction commenced and human remains were discovered, the project hit the skids. The project was later abandoned, and a judge awarded a large dollar settlement to the tribe.
- A local community put together a plan for a costly (over $30 million) interpretive center at the mouth of a major tributary to the Columbia River. Project proponents knew the preferred location was sensitive but thought that addressing the archaeological impacts would be sufficient for local tribes. When the project accepted federal funds a few years later, archaeological assessments expanded to include cultural concerns that local tribes had about the larger landscape. Five years later, the project proponents are still trying to get through the regulatory process so that construction can begin.
- Tribal monitors keeping watch over a U.S. federal agency's activities discovered that historic artifacts had been unearthed during an environmental contamination sampling project. What the agency archaeologists dismissed as a "waste site" turned out to be a landfill containing the garbage of workers who processed the plutonium for the world's first atomic bomb. While the agency refused to consider the landfill an important archaeological site, historical archaeologists mounted a national protest to the Keeper of the National Register to reverse the determination that the site was not important. The Keeper reversed the decision, and as of this writing, the agency is now conducting the archaeological mitigation.

These cases, some of which are discussed in greater detail in later chapters, illustrate how costs can be much more than financial. For the developer or agency involved, archaeological problems can lead to schedule delays, bad press, upset neighbors, and lawsuits. For the community and cultural groups who lose an important resource, these archaeological problems often lead to heartbreak, distress, and loss of quality of life in the community. For the archaeologists working for the developer or the government agency, archaeological problems can lead to lawsuits, damaged careers, and embarrassment. Nobody benefits from these events.

In this book, we provide a process for project managers to use for projects located in culturally sensitive areas around the world.

WHY ARCHAEOLOGY?

We target archaeology in this book because archaeological resources present a special set of problems for project managers. First and foremost, archaeological materials are typically buried and not easily seen from the surface. Despite advances in archaeological methods in recent years, one cannot always be certain about what lies underground. It is too costly to check for archaeological deposits at every location before construction begins, and thus some judgment must be used. Developing an adequate sampling strategy requires the expertise of professional archaeologists, geomorphologists, historians, and, if they exist, descendants of the people who may have lived at the site. All this takes time and money.

Complicating the situation is the fact that, if archaeological resources are found after construction begins, work stops while archaeologists conduct their subsurface excavations to assess the significance of the discovery. Removing artifacts and features can take considerable time and effort, and while construction activities may continue in other project areas, costs still accrue. Even after the archaeological digging stops, costs continue to mount as archaeologists begin the slow and expensive process of artifact description, analysis, and report writing.

In some areas of the world, cultural places such as archaeological sites and cemeteries are highly revered. This adds a spiritual and highly emotional dimension to already challenging technical, legal, and administrative situations. The solution may not always be as simple as removing the archaeological deposits or burials. As in the Port Angeles case described in Chapter 2, if the archaeological deposits are substantial enough, or the number of burials large enough, it may prove impossible to use the proposed location, regardless of how much money has been spent or how critical the project is to society.

The problems associated with the discovery and disturbance of archaeological resources are serious and can significantly affect the budget, schedule, and even successful completion of a project (Figure 1.1).

GOVERNMENTAL OVERSIGHT

Nations around the world are developing laws and procedures to protect their archaeological and cultural heritage. This is especially true in countries with large indigenous populations where the government must be responsive to indigenous rights. While the laws protecting archaeological resources vary across the globe,

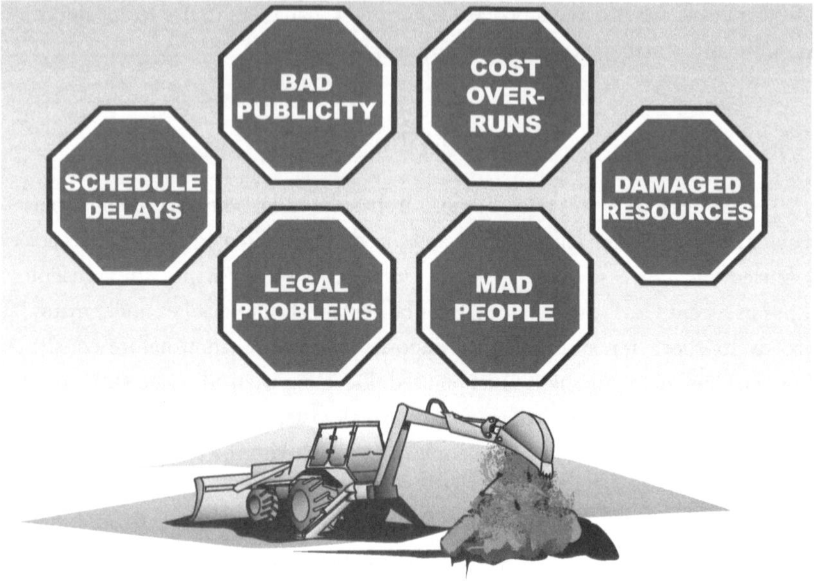

FIGURE 1.1. *Types of problems that can occur during the course of a project when archaeological resources are discovered during construction without a risk management approach.*

most involve three requirements: (1) to look for culturally sensitive resources prior to construction, (2) to consult with others to protect any important resources that may be found, and (3) to cease work if resources are discovered during construction while those involved decide what to do.

In some countries, such as the United States, the cultural protection system has become highly proceduralized. Laws have been passed, procedures established, and government agencies created to implement the system (ACHP 2009; King 2001, 2007, 2008). Other countries, especially those that are economically disadvantaged, have few established protocols or government staff to oversee development activities.

We argue that project managers need to take command of the situation regardless of the particular national regulatory requirements. Where requirements are highly proceduralized, archaeological studies will almost certainly be needed. A project manager needs to understand what must be done and why, so that the proper studies are completed before project schedules are affected. A project manager cannot simply rely on archaeological regulators to provide guidance. Regulating government agencies are notoriously understaffed and undertrained and may not always be able to give the best guidance.

Where requirements are less proceduralized or even nonexistent, there may not be anyone immediately available to advise a given project. Many countries have antiquities departments, but their efforts are usually focused on known monuments and heritage sites; thus they may not express much interest prior to construction, but will be interested if a discovery is made.

While many countries do not have well-defined requirements, international heritage groups have developed standards to assist in heritage protection. For example, the World Archaeological Congress, an international organization of archaeologists, promotes the training of professionals from economically disadvantaged nations and helps protect places of importance to countries, cultural groups, and communities around the world (World Archaeological Congress 2008). Similarly, the United Nations Educational, Scientific and Cultural Organization (UNESCO 2008) and the International Council on Monuments and Sites (ICOMOS 2008) have developed various charters and guidance documents focused on world heritage resources. For example, the ICOMOS *Charter for the Protection and Management of the Archaeological Heritage* (ICOMOS 1990) includes the following declaration:

> Development projects constitute one of the greatest physical threats to the archaeological heritage. A duty for developers to ensure that archaeological heritage impact studies are carried out before development schemes are implemented should therefore be embodied in appropriate legislation, with a stipulation that the costs of such studies are to be included in project costs. The principle should also be established in legislation that development schemes should be designed in such a way as to minimise their impact upon the archaeological heritage. [Article 3, Legislation and Economy, ICOMOS 1990]

Another example is the World Bank, an internationally supported bank that distributes more than a billion dollars each year for development projects around the world. The Bank supports development programs (e.g., bridges, roads, water systems, schools, hospitals) with the stated goal of reducing poverty in developing countries. It has developed procedures and guidelines for project developers to follow to ensure protection and enhancement of any cultural heritage their project may impact, which we have provided in Appendix A (World Bank 1994, 1999).

To prevent problems down the road, project managers should always contact the local or regional government agencies responsible for heritage preservation to identify specific requirements that will have to be met.

DEFINING ARCHAEOLOGY

For the purposes of this book, *archaeological resources* are broadly defined to include anything of antiquity that is found in the ground. Archaeological resources are found in a variety of contexts; they are most valuable when found *in situ*—that is, where they were originally deposited. Archaeological sites include the remains of cities, towns, villages, farms, buildings, homes, camps, work sites, and cemeteries (Figure 1.2). Any site can have any number of areas where soils contain artifacts and other clues about what happened there in the past.

While this book focuses on archaeology, it is important to remember that archaeological remains are part of a broader class of heritage resources (Table 1.1). Different peoples in different regions of the world have various names for these resources, such as *cultural resources, historical monuments*, or *historic sites*. In addition to underground archaeological remains, heritage resources also include historic buildings, places where historic events took place, places where traditional resources are found, spiritual places, and cemeteries.

In terms of importance, essentially anything from the past that is valued by the local community, a specific group, the state heritage agency, or professional archaeologists could be considered important. This underscores the need to work closely with the different parties to understand their issues, concerns, and expectations. The significance of any discovery will depend on where you are, who is in control, and whether anyone really cares. Particular attention should be given to human burials, because, as this book shows, projects that inadvertently disturb burials tend to be the high-profile cases with the biggest impacts.

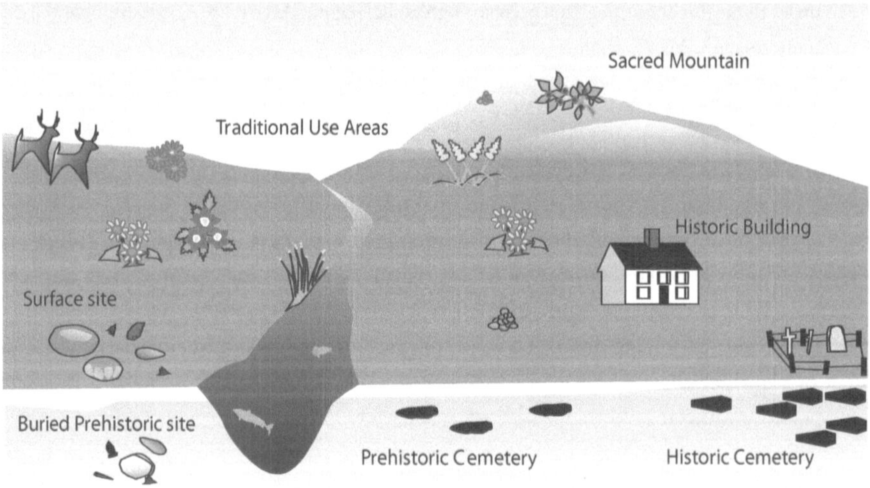

Figure 1.2. *An example of a cultural landscape showing the variety of heritage resources that are frequently encountered.*

TABLE 1.1. Selected types of cultural heritage sites

MAIN CATEGORIES	SUBTYPES	EXAMPLES	COMMENTS
Sacred sites	Burial sites	Xian, China; Tomb Fields, Bahrain	They are often discovered during the construction phase of projects.
Archaeological sites	Sites of religious or spiritual significance	Mecca, Saudi Arabia; Buddhist pilgrimage sites in Nepal	Important cultural sites are often inspired by religious beliefs and are still considered sacred places.
	Prehistoric sites	Mounds, middens, caves	These sites are often undetected or overlooked. They frequently shed light on use or overuse of natural resources, changing survival strategies, and social organization.
	Historic sites	Historic roads, bridges, dams, and other waterworks; fortifications; and walls	Many of these structures are still in use, such as Roman highways in Tunisia or hydrological features in Indonesia. They may also point to changes in sea level, vegetation, and hunting and agricultural practices.
	Engineering and industrial sites	Marib Dam, Yemen; The Great Wall, China; nineteenth-century industrial sites (e.g., train stations, ironworks, early textile mills)	The introduction of new technologies—metallurgy, mortars, arches and vaulting, industrial agriculture—can be documented and understood by studying artifacts and earlier structures. This in turn may suggest methods for conservation and may shed light on future avenues of technological advance. World Bank projects have been involved in conserving some of the world's greatest feats of engineering, including the Marib Dam in Yemen and The Great Wall of China.
	Submerged or marine sites	Ancient coastal settlements in the Mediterranean and Central America; sunken ships; wet sites.	New technologies of marine exploration have revealed many sunken ships and submerged sites of ancient human settlement.
	Sites with biologically diverse areas or protected reserves	Tikal, Guatemala; sacred groves in Ghana	Management policies that protect both cultural and natural resources should be developed.

Continued on next page

Table 1.1. Continued

Main Categories	Subtypes	Examples	Comments
Monumental sculpture	Cave sculpture	Chinese and Indian Buddhist cave sites	The protection of these sites depends on an understanding of the processes of deterioration that may affect them.
	Architectural sculpture	Thebes, Egypt; Petra, Jordan	Exterior sculpture is often damaged by polluted air and rising water tables.
Monumental painting	Cave or wall painting	Tombs in Luxor, Egypt; Tassili, Algeria	Conserving wall painting, in the face of large tourist flows, requires careful planning.
Architecture and town planning	Monumental architecture	Monte Alban, Mexico; Copan, Honduras; Wat Phu, Laos	Great works of architecture and urban planning demonstrate the introduction of new design principles and construction techniques.
	Indigenous or vernacular architecture	M'zab Valley, Algeria	Local materials, such as wood, mud brick, and stone, were used to build extraordinary architectural compositions.
	Historic settlements and town centers	Fez, Morocco; Quito, Ecuador	The protection of the historic core of cities depends on a comprehensive policy to address infrastructure and social needs.
Historic landscapes	Cultural landscapes	Cres, Croatia; land of the Dogon, Mali	Landscapes, whether designed, organically evolved, or relict, demonstrate human responses to changing environmental conditions.
	Historic parks and gardens	Sigiriya, Sri Lanka; Shalimar Gardens, Pakistan	Returning gardens to their original appearance may require research into plant materials.
	Trade routes and monuments and remains	The Silk Route from China to Europe; Pan-African trade routes; Mediterranean-wide commerce	Remains of ancient trade routes document early trade relations and cultural connections. Trading patterns, often long distance, are revealed in archaeological finds such as ceramics, metalwork, coins, or paleobotanical evidence.

Source: World Bank 1994.

AVOIDING A PROBLEM

So what can professionals do to avoid a project disaster when archaeological remains are discovered during construction? For a project manager, the answer is straightforward: become aware of the *possibility* of a problem and recognize the basic steps that should be taken *prior to starting construction*. We realize that project managers have many responsibilities and must coordinate a multitude of activities spanning numerous disciplines. Water permits, environmental permits, weekly cost and schedule reports, not to mention the day-to-day rigors of managing a complex construction project, are among the tasks at hand. The last thing a project manager wants is one more responsibility. In the twenty-first century, however, it is more important than ever for a project manager to become aware of archaeological issues as they relate to construction projects; indigenous peoples around the world are increasingly becoming more powerful, and one of the issues of greatest concern to many of these groups is protection of their cultural heritage.

A project manager need not become an expert in archaeology, any more than he or she must become expert in any of the other fields mentioned above. A project manager should have, however, enough working knowledge to interact effectively with archaeological professionals, regulators, and interested parties with whom he or she will be coordinating the necessary efforts (Figure 1.3). This book is designed to provide that working knowledge.

One simple and relatively inexpensive way to reduce the risk of an archaeological discovery is to commission a professional assessment. An archaeological assessment of a potential development is similar in many respects to an environmental assessment to identify contamination issues prior to buying a property. Individuals and corporations typically do not buy property unless they are certain that it is not contaminated with hazardous waste sites. How do they know? Standard due diligence involves extensive background research. What is the history of the property? Who owned it and what activities occurred there? What industrial processes were used, and where were chemical disposed?

Based on the background research, field inspections are made. Do any areas look suspicious? One might take soil samples to confirm the absence of contamination, but analyses are expensive, so a prospective buyer must use judgment on how much to spend given the results of the background research and field inspection.

So it is with archaeological concerns. If you want to know whether there could be archaeological resources in your project area, there are several actions to take: (1) do background research to understand what people did there historically; (2) visit the site to understand the geomorphology and examine the ground for archaeological

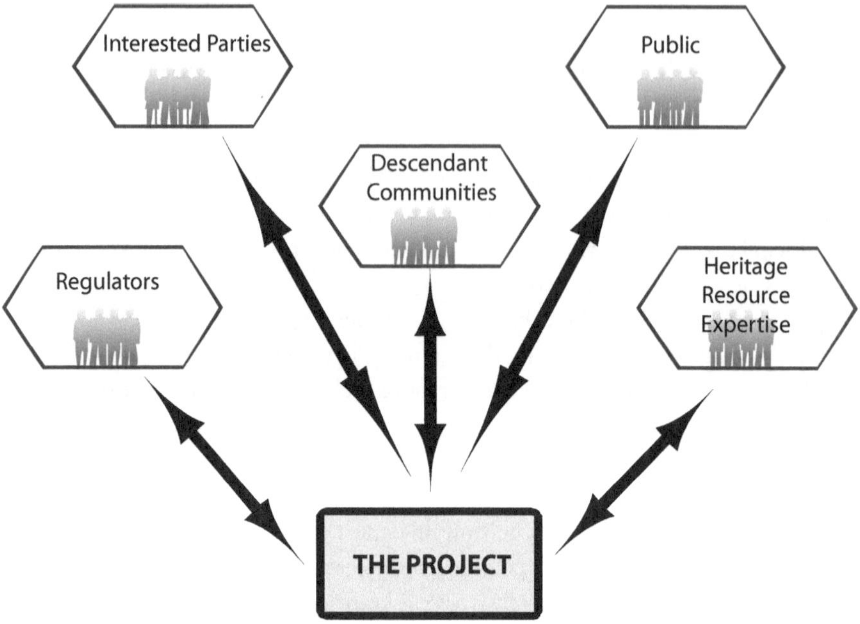

Figure 1.3. *Participants who need to be included in the heritage resource assessment process.*

evidence; (3) do some subsurface sampling to confirm the presence or absence of archaeological materials. The question is not really what to do, but rather how much to do to reduce risks to acceptable levels. This is a classic risk assessment problem.

Addressing risk is nothing new for construction projects. There are many kinds of problems that project managers must be prepared to handle during the life of a project. Suppliers may not deliver building materials when needed, the weather may not cooperate, subcontractors may not show up to do their work on time, archaeological permits or concurrences may not get issued, or there may be cost overruns. These risks must be managed to ensure that a project will be completed on time and within budget.

We have taken basic risk management principles and tools and adapted them to projects with archaeological resource concerns. By determining how likely it is that an important archaeological discovery will be made, and weighing that against the relative consequences if the discovery does occur, project managers can make better informed decisions concerning archaeological work that should be done during the course of the project. Actions that need to be taken to reduce risks can be documented with an archaeological risk management report and updated as necessary.

ORGANIZATION OF THIS BOOK

We begin by presenting two detailed case studies in Chapter 2 to give readers a sense of how the discovery of archaeological sites can affect projects located in archaeologically sensitive areas. These cases also provide a good introduction to basic archaeological concepts and compliance issues. In Chapter 3 we describe the types of actions that project managers can take to start managing an archaeological situation. Included are suggestions for consulting with interested groups, budgeting the efforts that will be needed, hiring archaeologists, and identifying regulatory requirements.

Chapter 4 describes the archaeological investigations than can be undertaken during the different project phases to keep risks under control. In Chapter 5, we present a risk management approach that project managers can use to evaluate the likelihood and consequences of archaeological scenarios facing the project. The risk management methods that we present, which will be well known to engineers and construction managers around the world, can help a project manager identify actions to avoid problems. In Chapter 6, we present three short case studies of archaeological disasters, which we use to demonstrate how a risk management approach might have changed the outcomes. In Chapter 7, we focus specifically on the challenges surrounding the discovery of human remains. We conclude in Chapter 8 with our ten principles of archaeological risk management. If a project manager can incorporate these principles into his or her project, the risk of turning the project into an archaeological disaster will be virtually eliminated.

Four appendixes are provided. Appendix A presents the guidance developed by the World Bank for addressing issues involving cultural property. Appendix B identifies several archaeological organizations recognized for their professional ethics and standards. Appendix C provides selected ethical codes from different heritage management organizations. Appendix D is a global guide to heritage management, including a country-by-country listing of procedures and governmental organizations responsible for heritage protection. A glossary of common archaeological and project management terms precedes the References at the end of the volume.

2 The Anatomy of Archaeological Disasters

Two high-profile cases from the Pacific Northwest of the United States provide a detailed look into the dynamic situations that arise when archaeological resources are discovered during a construction project. In the first case, a community began an expansion of its municipal sewage treatment plant with a loan from a federal agency. The cultural sensitivity of the area in and around the existing plant was well understood, and the project manager obtained archaeological expertise to help the project through the archaeological approval process. Archaeological sampling and data recovery were completed, and the project was approved to go forward, with construction monitoring by archaeologists. Soon after construction began, the heavy equipment operator encountered the ethnohistoric cemetery that many knew was in the area. About 60 full and partial graves were removed in hasty fashion, and no one notified the local tribe as required by a previous agreement with project proponents. When the local tribe discovered that the ancestral remains were being removed, tribal members called state officials and the project was shut down. The tribe was furious to learn that 20 boxes of bones had been transported to an out-of-state laboratory, which also violated the agreements regarding procedures for handling remains. The project was abandoned, million-dollar settlements were reached, and the community had to find a new location to construct its water treatment plant.

In the second case, a waterfront area was selected to prepare floating concrete pontoons for a major bridge repair project. A one-week archaeological assessment cleared the project, with a recommendation to monitor any construction that extended 4 feet below the surface. Soon after construction started, bone fragments and tools from a prehistoric cemetery and village appeared, and work quickly stopped. Millions of dollars were spent for cemetery removal and archaeological excavations, while project proponents, regulatory bureaucrats, and tribal leaders debated the fate of the project. The project location was eventually abandoned and a new one chosen. Monetary costs have been estimated at over $85 million.

We present these cases as classic examples of what can happen when basic principles of risk management are ignored.

THE BLAINE WASTEWATER TREATMENT FACILITY EXPANSION PROJECT

Blaine, Washington, is a small community north of Seattle near the Canadian border (Figure 2.1). In the mid-1990s, Blaine proposed to expand and upgrade its existing sewage treatment plant (Figure 2.2) to address expected population growth. Some knew that the existing plant was located near an important American Indian village

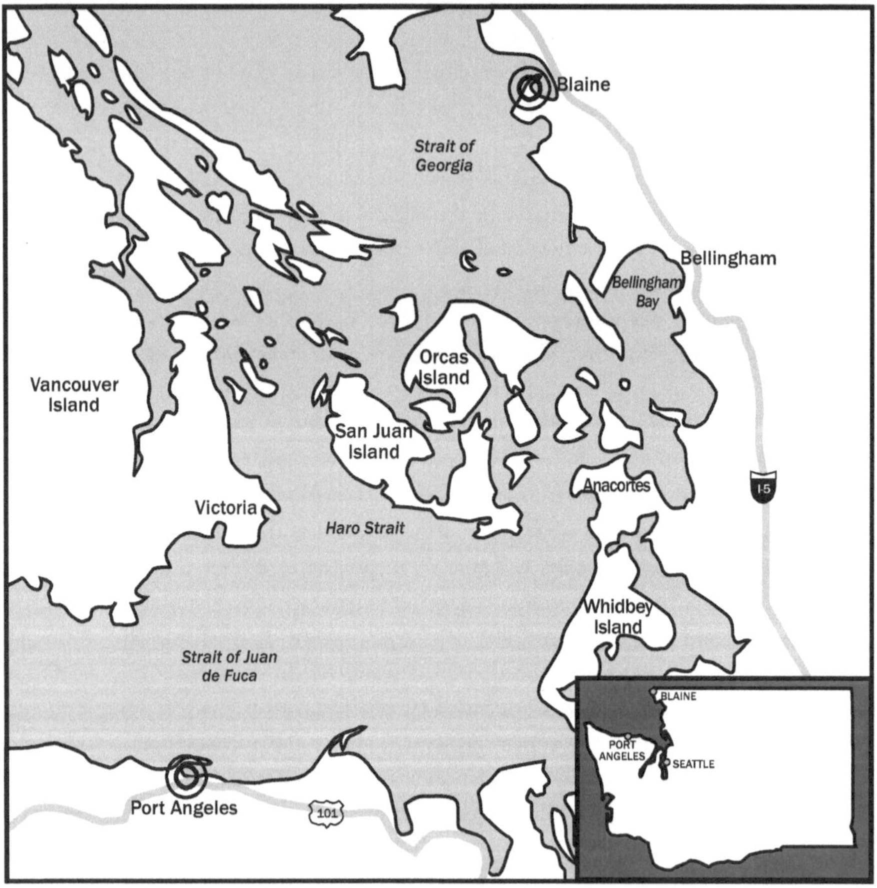

FIGURE 2.1. *Map of the Pacific Northwest showing locations of Blaine and Port Angeles, Washington.* Map by Faith Haney, Lithic Creative Media.

site. When the plant had been built in the 1970s, archaeologists had discovered archaeological deposits representing 4,000 years of use. Dozens of human burials also had been recovered. The site was also known to be important to The Lummi Nation, an American Indian group living adjacent to the City of Blaine.

Funding for the sewage plant expansion was provided in part by the Rural Development Agency (RDA), a federal agency that provides loans for this type of project. This federal funding required the agency to comply with various federal statutes, such as the National Environmental Policy Act (NEPA) and the National Historic Preservation Act (NHPA). Compliance with Section 106 of the NHPA would require the agency to evaluate the archaeological site for National Register of Historic Places eligibility. If the site was important enough to be eligible, the agency would be required to mitigate any adverse effects associated with the construction and operation of the plant. As is often the case in such projects, where the federal agency is only providing loans and not actually directing the project, the RDA authorized the project proponent, the City of Blaine, to handle the regulatory compliance.[1]

The Initial Assessment

To get an initial understanding of what they would need to do, the city hired a cultural resource consulting firm to conduct the initial assessment for approximately $10,000. The archaeological contractor, a well-known cultural resource firm from the Pacific Northwest, produced a report that described the archaeological site and cemetery, detailed the previous archaeological work that had been done, and provided recommendations concerning next steps. Importantly, the contractor provided critical information about the wastewater plant's construction decades earlier, when archaeologists from a local university discovered dozens of Native American burials. The archaeological contractor attempted to tie the previous excavations to the current plant location, and to assess the degree of disturbance of different site areas, but this proved difficult. Confirmation that the tribal cemetery had been used in recent times came from local tribal members, who reported that a fence once surrounded a burial area in this location, with graves marked with wooden crosses.

Defining the Regulatory Compliance Requirements

Once the project funding was secured, the city put out a bid for the next phase of archaeological work. The proposed scope of work called for the winning archaeological contractor to perform the National Register evaluation and coordinate with the various parties to define the archaeological work to be conducted. The details would be formalized in a memorandum of agreement, a legal document to be signed by the

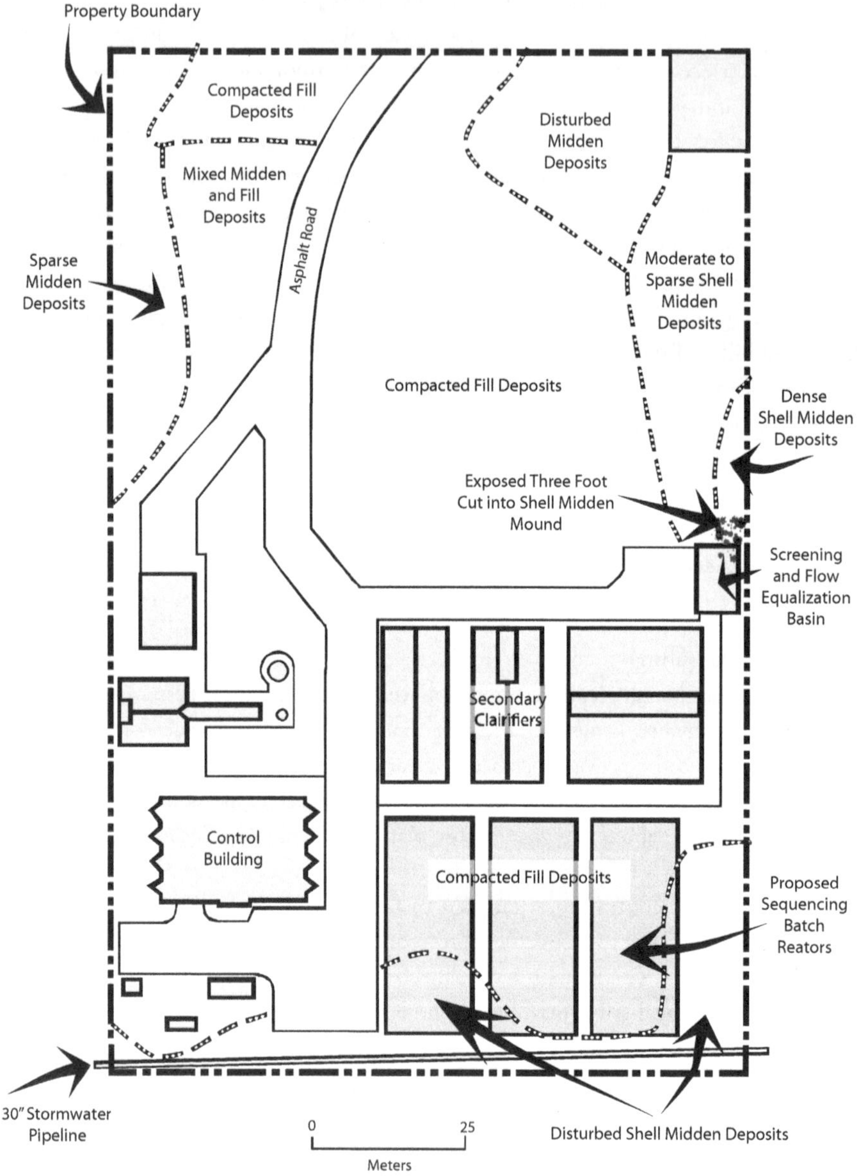

Figure 2.2. *Map of the Blaine Wastewater Treatment Expansion project area and the known locations of cultural deposits prior to the start of archaeological testing associated with the expansion project.*

city, the RDA, the Washington State Historic Preservation Officer (SHPO; pronounced "ship-o"), and the Lummi Nation. The winning contractor would then perform the work identified in the agreement.

Obtaining Expertise

The City of Blaine's request for proposal asked bidders to provide a fixed bid for the work. This forced bidders to predict the level of effort required to perform the archaeological work—something that could not possibly be accurately estimated beforehand. The archaeological consulting firm that had done the initial assessment, a local firm with a good reputation and extensive experience with similar types of projects, submitted a bid that was considered too high. The city selected a large international engineering firm with a small archaeological division, which had submitted a bid closer to what the city wanted to spend (approximately $50,000).

While the winning firm did have the lowest bid, several flags should have been raised by its qualifications. The firm's archaeologist responsible for coordinating the archaeological work was from another region of the country, had received a degree from Western Washington University, but had little experience in the project area, minimal experience working with American Indians from the area, and little experience working with human burials. The contract was to complete the National Register of Historic Places evaluation, consult with affected parties (i.e., The Lummi Nation), prepare a treatment plan (i.e., an archaeological work plan), submit a draft programmatic agreement (later called a memorandum of agreement) to affected parties, and then execute the treatment plan. The treatment plan would detail the work needed to recover archaeological information prior to construction (i.e., the mitigation). Archaeological monitoring during construction was also anticipated. The contract included about $15,000 for tribal consultation and $40,000 for the archaeological work. Additional funding for archaeological monitoring once construction began would be provided when construction details were known.

Developing the Agreement Documents

Upon award of the contract, the archaeological contractor began to plan the project. The project team was well aware of the cultural sensitivity of the project area and started consulting with the Washington SHPO to develop the work plan. Efforts were made to discuss the project with the local Lummi Nation, but little communication occurred.

The archaeological contractor and the state soon had the basics of the archaeological work plan identified, along with the procedures to follow given possible situations, such as the discovery of human remains. The parties agreed that the archaeological site was eligible for listing in the National Register of Historic Places and that the project would have an adverse effect on the site.

A memorandum of agreement was negotiated, defining the requirements to be met by the archaeological work in order for the plant expansion to occur. The document was signed by the RDA, the Washington SHPO, and Advisory Council on Historic Preservation, and, at the last minute, the Lummi Nation. The key parts of the final agreement were:

- preparation and implementation of a work plan for the pre-construction archaeological recovery work (detailed in the treatment plan);
- preparation and implementation of a site protection plan, calling for monitoring during construction, regular reporting, and actions to be taken if new archaeological deposits were discovered during construction;
- guidelines for curation of artifacts;
- tribal protocols for handling of Native American human remains;
- professional qualifications of archaeological consultants, citing the Secretary of Interior's Professional Qualifications/Standards for Archaeologists (Federal Register 48 [190]:44738-44739);
- dispute resolution plans outlining the process to be used if any of the parties could not agree on a particular action.

The treatment plan, included as an attachment to the memorandum of agreement, was a substantive document over 40 pages in length. It provided a description of the expansion project, the environmental setting, and the cultural setting, extracted mainly from the initial assessment report. Research issues to be addressed during the archaeological investigations were also identified. The primary research goal was to learn about the structures that prehistoric people had lived in at the site. Another key goal was to work with the Lummi Nation to develop expeditious and culturally sensitive procedures for removing any human remains that might be discovered.

The treatment plan contained methods to be used throughout the project. Methods included nondestructive technologies such as ground-penetrating radar to develop a subsurface map of the project area, trenching to provide stratigraphic information, and twenty 1 × 1 meter hand-excavated test units. Key points in each section that bear on subsequent events included:

- Phase 1, Consultation: Consultation with the tribe was recognized as paramount because the site was clearly important to the tribe and because the project area likely contained more human burials. The plan was to approach consultation in three steps: introductory meetings, discussions, and then collaboration.

- Phase 2, Field Investigations: Using information gained from the earlier geophysical investigations, archaeological investigations would involve two aspects. First, two intersecting backhoe trenches would be excavated to clarify the disturbed and undisturbed deposits. Backhoe excavations would be monitored by at least one professional archaeologist, and stratigraphic profiles along one wall of each trench would be drawn and photographed. Second, as many as twenty 1 × 1 meter units would be excavated where the backhoe trenching results had indicated manual excavations would be most productive. Methods would involve standard archaeological techniques, such as troweling to maintain strict horizontal and vertical control; shovels would be used in areas with few artifacts. In the event that human remains were encountered during the backhoe or hand excavations, all work would stop; the RDA, the Lummi Nation, and the Washington SHPO would be notified, and the agreed-upon procedures would be implemented.
- Phase 3, Construction Monitoring: The monitoring was intended to ensure that significant cultural materials were recognized and properly documented if encountered during construction. If cultural materials were identified, construction work in the area would cease and the find exposed and mapped in plan view by the archaeologist. Features and structural remains would be bisected, one-half removed, and sketched and photographed. If the find consisted of human remains of any sort, the agency, tribe, and city would be contacted and the remains treated according to protocol.
- Phase 4, Laboratory Analysis: This section contained brief information about how artifacts would be analyzed. Human remains would be documented in place and then removed to a nearby storage facility.
- Phase 5, Reporting: Archaeological investigations would be documented in a professional report.
- Site Protection Plan: The Site Protection Plan directed the city to obtain the services of a professional archaeologist whenever ground-disturbing activities were needed. If human remains were ever encountered, work would stop and the city would contact the designated tribal representative.
- Curation: The plan was to curate collections and records at a local university. If any human remains were found, their disposition would be determined upon consultation between the agency and the tribe.

Pre-Construction Mitigation

Before construction could even begin, the archaeological mitigation work needed to be performed. The treatment plan first called for two long trenches to be excavated with heavy equipment so that the archaeologist could see where the undisturbed artifact-bearing deposits were located. The archaeologist would use the information gained from the two trenches to locate the twenty 1 × 1 meter excavation units.

On the first day of trenching, the crew identified a bone as human. The archaeologist collected the bone and contacted the tribe, following the procedures identified in the memorandum of agreement. The archaeologist met with the tribal representative, the bone was transferred, and protocols for future discoveries were agreed to. One protocol called for a summary to be faxed to the tribe each day. No discoveries of human remains were reported during the remainder of the mitigation.

Over the next 10 days, several trained archaeologists and two tribal representatives completed the mitigation at the construction site (Figure 2.3). Many artifacts were recovered, but no obvious information about living structures was recovered. No more human bones were recovered. Based on the results, the archaeologist recommended that the city be allowed to begin construction. The schedule did not allow time for preparing an archaeological report of the material discovered during the mitigation work. Soon thereafter, site preparation activities began, marking the next phase of archaeological work: construction monitoring.

Construction Monitoring

A two-person team, supervised by the lead archaeologist, was on hand to monitor site preparation activities. Archaeological monitors would observe heavy equipment excavations and look for archaeological artifacts and bones. Two tribal monitors were supposed to be part of the team, but tribal participation in the monitoring was suspended on account of budget problems.

Bones were discovered from the very start of construction, but none were positively identified as human by the crew (who had little expertise in identifying human bones). On the fourth day, more bones were recovered, some of which were positively identified as human. Work continued, and for reasons that have never been explained, no one contacted the tribe or the state, as was agreed to in the memorandum of agreement and treatment plan.

On Monday, the archaeologists continued to find human bones. As called for in the procedures, work ceased in the area while the archaeologists removed the bones and associated deposits and made notes. On the sixth day, the city started expressing

(FACING PAGE): **FIGURE 2.3.** *Map of the Blaine Wastewater Treatment Expansion project area, showing locations of archaeological test trenches and excavation units, the archaeological mitigation units, and the area where human remains were later discovered during construction. A sampling plan that was focused on the question of whether human remains were located in the project area would likely have located the burials prior to construction.*

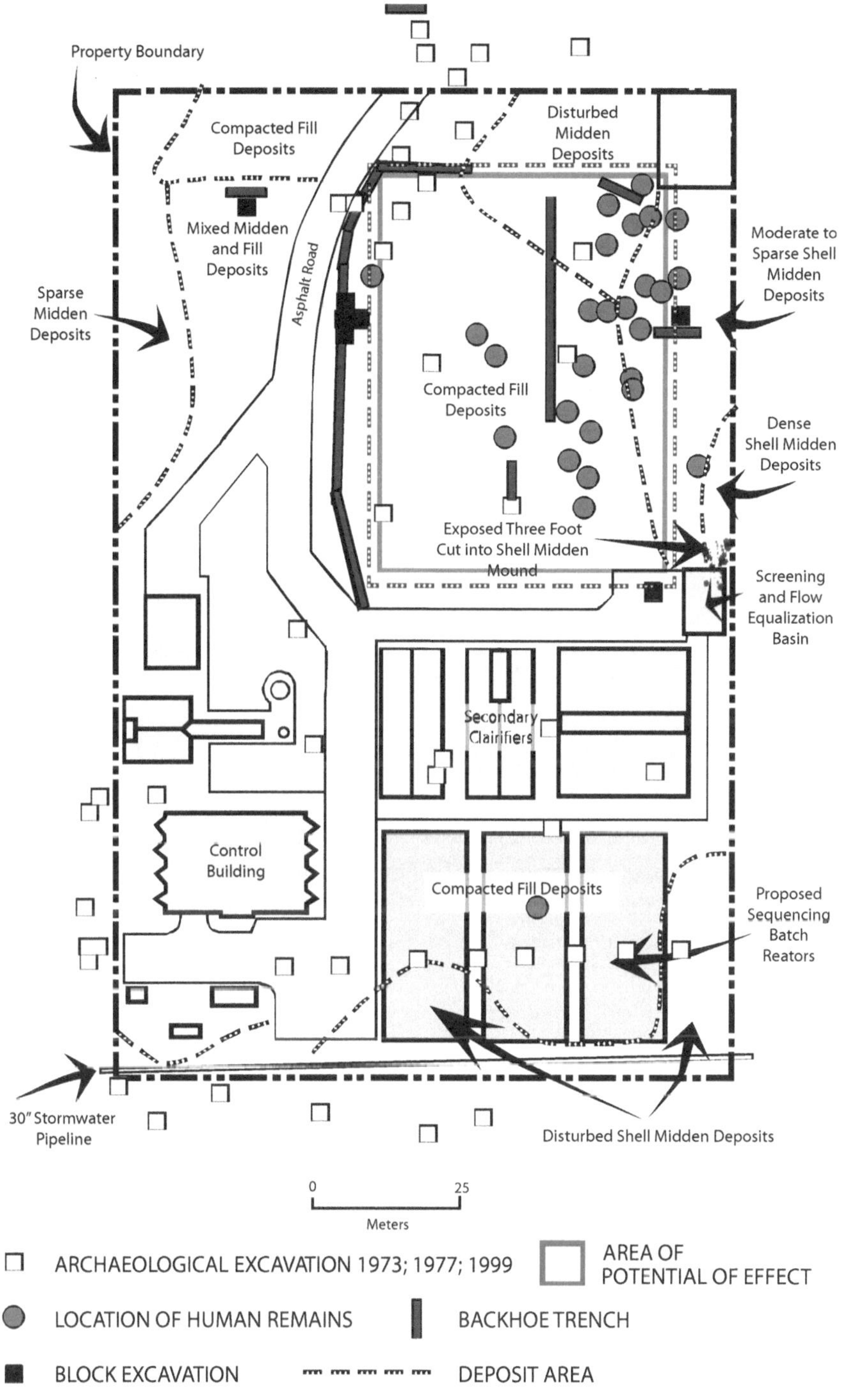

Property Boundary
Compacted Fill Deposits
Disturbed Midden Deposits
Mixed Midden and Fill Deposits
Asphalt Road
Moderate to Sparse Shell Midden Deposits
Sparse Midden Deposits
Compacted Fill Deposits
Dense Shell Midden Deposits
Exposed Three Foot Cut into Shell Midden Mound
Screening and Flow Equalization Basin
Secondary Clairifiers
Control Building
Compacted Fill Deposits
Proposed Sequencing Batch Reators
30" Stormwater Pipeline
Disturbed Shell Midden Deposits
0
25
Meters
ARCHAEOLOGICAL EXCAVATION 1973; 1977; 1999
AREA OF POTENTIAL OF EFFECT
LOCATION OF HUMAN REMAINS
BACKHOE TRENCH
BLOCK EXCAVATION
DEPOSIT AREA

concerns about the costs associated with the construction stoppages that were occurring while the archaeologists collected the bones and associated archaeological deposits. The lead archaeologist agreed to scale back from two archaeologists to one archaeologist when little or no digging was going on, but for the main excavation, two archaeologists would still be needed because they were in the middle of such a rich archaeological deposit. Bones and prehistoric artifacts continued to be found through the tenth day, still with no notification to the tribe or the state.

Major construction excavation began on the eleventh day. That day, six burials were unearthed and recorded. The next day, three more burials were located, one of which appeared to be intact. The next day six more burials were unearthed. The bones were boxed up into six boxes and taken away. On fourteenth day, bones from 11 different burials were recovered. The questions begs, "Were they coming from the cemetery that tribal elders had said was in the area?"

By the end of the second week, 30 burials had been identified, removed, and placed in paper bags and cardboard boxes. Except for a voice mail left at the tribal headquarters, no attempts were made to reach the tribal contact, the state, or any other signatories of the memorandum of agreement, as mandated in the legal agreement documents.

Project Shutdown

On the sixteenth day, a Lummi representative came by the project and learned about the many human remains that had been collected and transported offsite. Lummi representatives reported the situation to the Washington SHPO, and within days, work at the construction site was stopped. The SHPO terminated the memorandum of agreement immediately, stipulating that the removal of over two dozen burials was in violation of the agreement document. The lawsuits began.

The Lummi arranged to take possession of the remains as soon as possible and hired its own archaeologists to determine exactly what had happened. It was learned soon afterward that a large amount of soil from one area of the project site had been excavated and trucked to a local landfill where it was spread for road cover and other purposes. The archaeologists had not investigated that specific project area, and it turned out that the area had contained many graves. The skeletal material had been smashed and broken up and then transported offsite and spread over large areas. Ten years later, the recovery effort is ongoing and fragments of bones continue to be recovered by tribal archaeologists and volunteers—at a staggering cost to the project principals.

The lawsuits took years to resolve. Reported settlements of $5 million to the tribe, from both the engineering firm that had provided the archaeological expertise and from the city, were made. Other settlements were made but not disclosed. The expansion project never started up again, and ultimately, a new location was found for a new wastewater treatment facility. Tribal members were devastated by the knowledge that their ancestors had been unearthed, desecrated, and spread across the landscape. Remains continue to be recovered and reburied by the Lummi. No archaeological report was ever produced. By any measure and from any perspective, this project was an archaeological disaster.

Observations

The events at Blaine point to a number of weaknesses in the process that contributed to its overall failure. These are grouped into the following categories:

- archaeological compliance;
- tribal consultation process;
- contractor funding and qualifications;
- following agreed-to procedures.

Archaeological Compliance

Despite the known presence of a Native American cemetery in the immediate vicinity of the project area, and the recognized sensitivity that burials had to the neighboring Indian tribe, the archaeological actions developed and approved through the archaeological compliance process failed to protect an important cemetery. The expansion project probably should never have progressed once the cultural sensitivity of the area was demonstrated. Since the project did progress, the negotiating parties should have put more emphasis on characterizing the project area and developing a sampling plan that would find burials if they existed in the construction zones (Figure 2.3). The project should not have relied on archaeological monitoring, given the lack of subsurface information for particular construction areas. Despite the efforts of the city, the federal agency, the state archaeology office, and the archaeological contractor, the end result was utter failure. If this teaches us anything, it is that complying with well-defined archaeological compliance processes alone is no guarantee that problems will be avoided.

Tribal Consultation

The archaeological contractor clearly had trouble getting the Lummi Nation to participate in the project planning and execution. One of the things that strikes us about this case is the delegation of the consultation requirements to the contractor. As the federal agency, the RDA, had the legal responsibility to consult, but because of its minor role in the project, it had no onsite presence and its consultation efforts involved little more than letter writing.

From a community perspective, and given the cultural sensitivity of the project area and the damage that had been done to the cemetery when the plant was built in the 1970s, we would have expected city leaders to have initiated formal government-to-government meetings with Lummi leaders to discuss the project. Sending an archaeological contractor in the absence of any formal consultations was a mistake.

When the contractor failed to get tribal response to any of the agreement drafts that had been faxed or mailed to the tribal offices, the city should have realized that something was wrong and needed to be fixed.

Certainly the outcome might have been different if meaningful consultation had occurred. We could point to the archaeologist's lack of experience in tribal consultation as a reason that consultations were not more effective. It is unreasonable, however, to think that an archaeologist can conduct such sensitive consultations. The city should have been consulting with the tribe, and the fact that they did not indicates that relations between the city and tribes were dysfunctional.

Admittedly, if a construction project is going to result in the destruction of someone's cultural heritage, consultation with that party will be inherently difficult. It is not at all uncommon for tribes or other cultural groups to be less than cooperative with agencies and contractors during a project. They have little incentive to help a project construct something that has little if any benefit to them and in fact is doing harm to their heritage.

Nevertheless, the agencies, city, and contractor did understand the sensitive nature of the project; they committed to notify the tribe and other signatories if human remains were encountered. Better consultation might have yielded better results, but the fact remains that consultation did result in a process of notification, which, if followed by the archaeologist, would have stopped construction soon after the burials started being unearthed.

Although the consultations did not achieve the hoped-for success, project personnel did gain a sense of the values the tribe wanted instilled in the project, which were in fact used in developing the project's policy concerning the treatment of human remains. The parties understood that:

- site preservation should be the primary goal;
- fragmentary and intact human remains should be considered equally sensitive;
- though the ground was disturbed in places, the presence of fragmentary remains, even mixed with imported sediments, rendered those sediments as sensitive and, perhaps sacred, ground;
- certain tribal people are trained to work with human remains, and they should be consulted with regard to the handling of any burials or fragmentary remains located on the site.

Still, without meaningful tribal involvement and participation, words such as these tend to lose their meaning and importance. Had there been meaningful involvement, we suspect that the protection of human remains would have taken higher priority and the outcome at Blaine might have been far different.

Contractor Funding and Qualifications

The archaeological team fielded by the archaeological contractor was underqualified and underfunded for the type of archaeological work that was required to adequately protect the resources. Cost concerns drove the city to go with the lowest bid, even though the lowest bidder was from out of the area and had little experience in Northwest Coast archaeology. The lead archaeologist, for example, did not have the expertise and experience required to successfully complete this project. He had conducted little work in the region, was not familiar with current archaeological research in the area, and had little, if any, experience in working with Native Americans.

Cost constraints kept the archaeological contractor from obtaining the osteological expertise that was needed to identify and handle human remains. If someone on the project team had possessed the specialized skills for distinguishing human from non-human bone, over 70 human bones might not have been misidentified in the field during the archaeological mitigation. If these human bones had been identified as such, those involved in the project may have altered their assumptions and considered additional mitigation prior to construction.

For example, in one document, the archaeologist addressed the budget implications of hiring someone with osteological expertise, noting that finding a subcontractor to undertake the analyses would seriously impact the budget, especially if numerous human remains were encountered.

Another casualty of the budget was the plan to have tribal monitors observing when construction began; tribal monitors were dismissed when the city could not afford to pay them. Even the archaeological monitoring was scaled back. When burials started to be unearthed, there was no one with expertise available to carefully document and remove the remains. Burials were removed in about 30 minutes each, a surprisingly short amount of time for such a delicate operation.

Budget constraints and pressure to reduce costs are nothing new in the heritage resource management business. Most companies, agencies, and municipalities who are forced to pay for contract archaeology always seem to lack enough funds. It is a way of life for most archaeological contractors. Clients will only set aside so much money in their budget for archaeologists; where those numbers come from is often a mystery.

It is the resources needing protection, however, that should determine how much work needs to be done, not the available budget. It is as if the city told the contractor how much money was available for the archaeological work and the contractor designed the work accordingly.

We saw nothing in the documentary record to suggest that the contractor ever explained the costs and risks to the city. The archaeological contractor knew this was a very sensitive project, given the fact that graves had been disturbed and/or destroyed during the initial construction efforts. Ideally, the contractor should have communicated something like the following to the city:

> The budget we propose is sufficient for completing the proposed tasks. The city should realize, however, that this is a very sensitive location and may contain extensive cultural deposits. If, during the mitigation phase, we encounter such a scenario, it is likely that the signatories to the memorandum of agreement will require more data recovery, which will require more funding and delay start of pre-loading. Similarly, once construction begins, if the archaeological monitors discover additional human remains or significant cultural features, disposition of these remains will also likely impact cost and schedule. The tribe will likely have additional concerns and requests that will need to be addressed before construction can continue.

Every project has risks, and project managers constantly weigh risks against costs as part of their decision-making process. Successful project managers tend to be better at risk management than unsuccessful ones. But central to good risk management is having all the information. It is incumbent on professional archaeologists to inform their clients of the risks of inadvertent and unanticipated discoveries, and to design their archaeological sampling to further reduce those risks. In the Blaine case, we found no documentation that the archaeologist explained the risks of underfunding to the client.

Following Agreed-to Procedures

Despite clearly delineated procedures for dealing with any encountered human remains, these were ignored. By the time the interested parties found out about the discovery, a Lummi cemetery had been destroyed. If work had stopped when the first human remains were discovered during construction, and if the proper notifications had been made, much of the cemetery destruction could have been prevented.

The memorandum of agreement and the treatment plan had been prepared specifically to protect important resources and human remains at the ethnohistoric village and cemetery. It was clearly stated in several places that, in the case of inadvertent discovery of Native American burials or human remains during construction, archaeological fieldwork, or laboratory analysis, the city would notify the tribe and consult with them over the handling and disposition of remains.

The archaeologist wrote the agreement documents so he clearly understood the commitment that had been made. Despite his familiarity, he did not follow the mandated process mandated in the documents, nor did he follow the protocol established with the tribe during the mitigation phase: when a human bone is discovered, the tribe is notified. The bewildering lack of notification, especially when intact burials began appearing, the failure to stop work when construction crews clearly entered an area with numerous human remains (i.e., a cemetery), and the use of unqualified people to remove intact human remains all suggest serious inadequacies with regard to the archaeological services provided.

The Port Angeles Graving Dock Project

In 1998, bridge inspectors in Washington state reported that the eastern portion of the Hood Canal Bridge, a heavily traveled bridge spanning the Straits of Juan De Fuca, was approaching the end of its design life. Efforts to obtain the estimated $200 million in funding for the repair work commenced, the project was included in the state budget cycle for 2005, and a schedule for completion in 2007 was developed. A key part of the project involved finding a location to fabricate 19 large concrete pontoons, each weighing 9 tons, which would be needed to support the new portion of the floating bridge.

By 2002, concern about the project started to surface because a location for the ancillary facility to make the pontoons had not been found. The plan was to make each pontoon in a dry dock (called a *graving dock*) and then to float the finished pontoons to the bridge location. About 20 acres of waterfront property were needed for the graving dock project. Recent experiences showed that meeting regulatory approval, especially those requirements involving endangered salmon species, would take time. To help meet the complex requirements, the state put the project on an "accelerated" basis to streamline the regulatory process and keep the schedule on track.

Securing a location for the graving dock proved difficult for the Washington State Department of Transportation (WSDOT), the state agency assigned responsibility by the U.S. Federal Highways Administration (FHWA) for repairing the bridge and completing the regulatory compliance requirements. By November 2002, selection of a location had become critical to the overall bridge project schedule. At the last minute, the City of Port Angeles, Washington, offered a 22.5-acre parcel of waterfront land (Figure 2.4) that had been the site of a lumber mill since the early 1900s. WSDOT accepted the Port Angeles proposal, and the challenge quickly shifted to getting various environmental and other permits so that construction could begin.

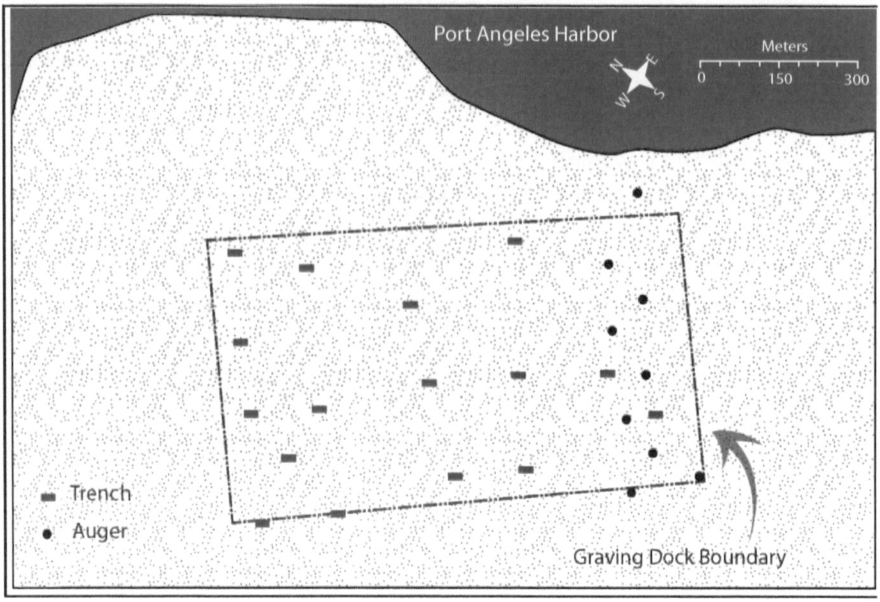

FIGURE 2.4. *Map of the Port Angeles Graving Dock project area, showing the Graving Dock boundary (area of potential effect), trenches, and auger holes investigated for the initial characterization effort.*

Regulatory Compliance and Site Characterization

Archaeological compliance was one of several regulatory reviews that needed to get done for the Port Angeles project. However, because the agency regulatory staff believed that the biggest challenges to their schedule would be posed by the U.S. Endangered Species Act and the National Environmental Policy Act (NEPA), they paid little attention to cultural issues. Cultural and archaeological concerns should be addressed under NEPA, but often they are not given serious consideration. WSDOT, Olympic Region, focused its cultural and archaeological efforts on the National Historic Preservation Act Section 106 compliance process, in isolation from the environmental team that was pursuing the environmental regulatory issues.

The land offered to the state for the graving dock was well known as an important area associated with the local Lower Elwha Klallam Tribe. The Klallam people had been living at the proposed location until the 1900s, when they were forced out of the area. In 1915, during construction of the lumber mill, a cemetery was discovered and subsequently destroyed during site clearing. At the time, newspapers reported that backdirt containing broken pieces of human bone was used to backfill the various trenches and low spots on this ancient beachfront landform.

In 1995, the City of Port Angeles had an established relationship with the Lower Elwha Klallam, whose tribal headquarters is located several miles inland. The city recognized the cultural sensitivity of the waterfront when it created the Shoreline Master Plan Regulations of the City of Port Angeles. That plan stated the following:

Applicability

- Archaeological and historic resources, because of their finite nature, are valuable links to our past and should be considered whenever a development is proposed along the state's shoreline. Where such resources are either recorded at the State Historic Preservation Office and/or with local jurisdictions, or have been inadvertently uncovered, the following policy and regulations apply:

Policies

- Due to the limited and irreplaceable nature of the resource, public or private uses and activities should be prevented from destroying or damaging any site having historic, cultural, scientific or educational value as identified by the appropriate authorities.
- Due to the traditional villages, homesites and burial grounds of the Lower Elwah S'Klallam [sic] Tribe documented along the Port Angeles harbor, the vested interest of the Tribe should be respected.
- The City of Port Angeles will inventory and map cultural sites with consideration of the cultural significance and protect sensitive information from public disclosure. *[Reproduced from WSDOT Evaluation Document, footnote from Section 3, p. 30.]*

Despite this apparent regard for the tribe, however, just seven years after the Shoreline Management Plan was issued, the city sold to WSDOT a 22.5-acre piece of extremely culturally sensitive land on the shoreline for the graving dock project. This would prove to be a mistake, as over a 24-month period, WSDOT would disrupt 276 full burials and 500 partial burials of Lower Elwha Klallam ancestors. The costs for recovering the burials and the millions of artifacts from the ancient village were staggering.

For our purposes—examining the mismanagement that created this archaeological disaster—the key part of the story began with the initial assessment of the location, which was launched soon after the Port Angeles location was offered and chosen. WSDOT had no professional staff within the agency to conduct the cultural assessment, so it subcontracted the work. On October 21, 2002, WSDOT sent a task order to an archaeological contractor with which it had a contractual relationship (an "on-call contract"). The scope of work, to be completed by November 15, 2002, asked for the following:

- literature and archival scope and review;
- ground survey;
- structural review;
- tribal coordination;
- summary report.

The project area to be assessed, referred to as the *area of potential effect*, "would include all the new ground disturbance plus any areas of concern noted by the appropriate tribe." WSDOT added, "The project is being fast-tracked and we encourage you to accomplish the task with urgency."

Finalizing the contract required more time than anticipated, and WSDOT and the archaeological contractor were not able to come to terms until November 13, 2002. The contractor then conducted the work. The summary report and recommendation for regulatory compliance reported the following:

- The area of potential effect was the 9.55 acres defined as the graving dock._
- Background research identified that an important Indian village, named *Tse-whit-zen* (pronounced Che-Weet-Son), was located nearby, and that the *Tse-whit-zen* cemetery was in the general vicinity of the mill complex that had occupied the site since 1915, which had disturbed hundreds of Indian bones when ground was broken for the foundation. It was further noted that cemeteries were typically near the beach, fronting the water.
- Tribal coordination consisted of a telephone call and a brief meeting with a neighboring tribal contact.
- Archaeological testing included 9 auger holes to a depth of 10 meters and 17 trenches approximately 2 × 1 × 2 meters deep. Trenches were placed on a 40-meter grid, with some flexibility for working around concrete-covered areas. No evidence of significant historic or prehistoric archaeological resources was found. Some stratigraphic information was contributed, and industrial artifacts unearthed suggested that the upper 1.6 meters below the surface consisted of disturbed deposits associated with the last 100 years of mill construction and operation. The estimated cost of the fieldwork was $6,300 (WSDOT provided backhoe services). Bad weather and equipment difficulties resulted in three days of work, rather than the planned five days.
- In the archaeological contractor's judgment, there was only a low probability that buried archaeological resources existed in the project area.Nevertheless, the contractor recommended archaeological monitoring in construction areas where excavation would exceed 4 feet (1.3 meters).

The findings and recommendations made in the assessment were sent to the Washington SHPO and the local tribe, and both parties concurred. The tribe added that work should proceed with caution. WSDOT prepared a monitoring plan and distributed it to the SHPO and tribes.

Initial Construction Phase

In late February, several weeks behind schedule, WSDOT released a bid package to construction firms to build the graving dock facility. The bid included a 90 percent design rather than the more normal 100 percent design, in order to stay on schedule. The procurement took longer than expected, however. WSDOT opened the bids on June 18, 2003, two months later than originally planned, and the contract was awarded to the low bidder on June 24, 2003.

The construction contractor started mobilizing soon thereafter. Groundbreaking for the Port Angeles Graving Dock was held on August 3, 2003, and site clearing commenced. Buildings were torn down, the site was leveled, concrete slabs covering large portions of the site were broken up and removed, and concrete foundations of buildings were pulled out. Two weeks later, on August 16, 2003, a WSDOT supervisor observed dark soil and clamshells where a building foundation was being removed. Archaeologists from the archaeological firm, the local Lower Elwha Klallam Tribe, and the U.S. Army Corps of Engineers (the Corps) (involved because the project needed a Corps permit) were called to the site and confirmed that undisturbed archaeological deposits had been unearthed. Work in the immediate area ceased, while site-clearing efforts continued in the remainder of the project area.

Additional archaeological discoveries and human remains fragments, referred to as "inadvertent discoveries," were made over the next week. The various archaeological representatives grappled with what to do next. Finally, on August 26, after discussions with the tribe, the WSDOT secretary ordered all work to stop until additional archaeological characterization could be done. Human remains that had been recovered were repatriated to the tribe the next day.

United States cultural protection procedures, referred to commonly as Section 106 of the National Historic Preservation Act, provide fairly detailed procedures in the event of an inadvertent discovery such as this. The parties involved—the developer (in this case, Federal Highways Administration and WSDOT), the State Historic Preservation Office (in this case, the Washington Department of Archaeology and Historic Preservation), and any tribes or interested parties (in this case, the Lower Elwha Klallam Tribe)—were required to get together and work out an agreement, with a detailed plan, to protect the resource or mitigate project impacts. In this case, WSDOT, the FHWA, the SHPO and the tribe spent the next seven months developing a formal memorandum of agreement. To facilitate tribal involvement, WSDOT provided $80,000 to the tribe to help defray current and upcoming expenses.

Second Archaeological Characterization Effort

The negotiation process required the parties to plan for additional site characterization and archaeological assessment. With the site cleared, it would be possible to do more sampling and get a better understanding of the area. With this new data, archaeologists created three-dimensional maps of cultural deposits. The memorandum of agreement was signed September 18, 2003. It called for two additional weeks of archaeological work, estimated to cost $40,000. Work started the day the agreement was signed.

Over the next two weeks, the archaeological contractor mechanically excavated 80 trenches. During this second phase of testing, the archaeological contracting firm was able to identify four areas of archaeological interest (designated as Areas A, B, C, D; Figure 2.5). Based on the information collected, the contractor determined that the archaeological site should be eligible for listing on the National Register of Historic Places, meeting the criterion of being able to contribute important prehistoric information.

The site was described as a prehistoric site, an ethnohistoric site, and a cemetery and was identified as part of the village of *Tse-whit-zen*. The SHPO concurred that the site was eligible on October 10, 2003, opening the door to additional archaeological work (considered "data recovery") to mitigate the effects of constructing the graving dock on the archaeological remains. The results of the second phase of archaeological characterization were released in a one-page memo soon after completion of the work. Three-dimensional models of the cultural deposits identified the locations and dimensions of archaeological deposits located within the project area.

With funding from WSDOT, the Lower Elwha Klallam had hired its own archaeological consultant to observe the second phase of archaeological work. The tribe's consultant offered a different interpretation of the data recovered in the 80 trenches and a different assessment of the cultural and archaeological significance of the deposits (Table 2.1). Where the WSDOT contractor saw culturally sterile, disturbed deposits with little or no value to the tribe or the archaeological community, the tribal consultant saw a rare, information-rich, one-of-a-kind, irreplaceable time capsule for learning about the past which did, in fact, have special meaning to the tribe. The tribal consultant turned out to be correct.

Negotiating Agreement Documents

Based on the information collected during the archaeological work and the construction needs of the project, the two federal agencies, two state agencies, and the Lower Elwha

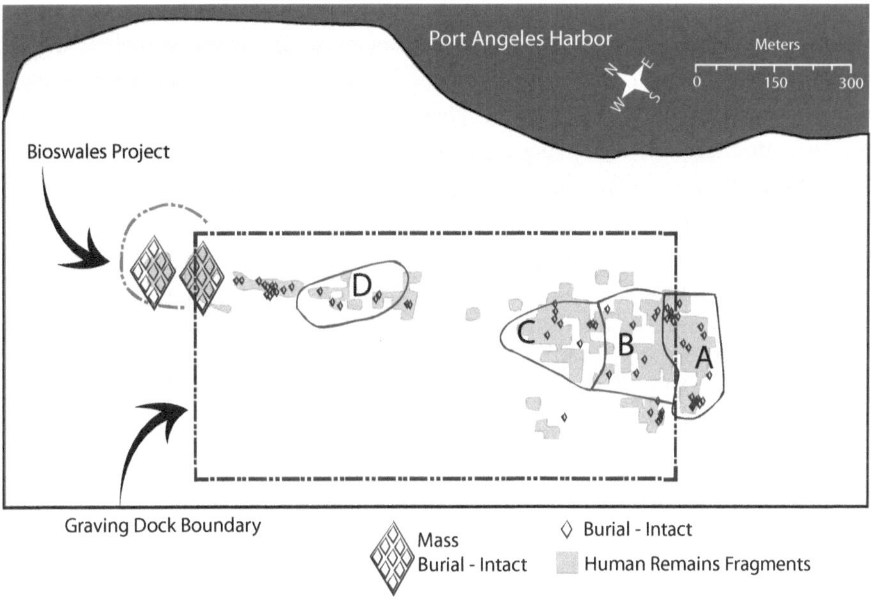

FIGURE 2.5. *Map of the Port Angeles Graving Dock project area showing archaeological areas (A, B, C, D, and Bioswales locations) identified during the second characterization phase.*

Table 2.1. Estimates of undisturbed cultural material made by the WSDOT archaeological contractor who performed the work, and the tribal consultant observing the work

AREA	WSDOT CONTRACTOR	TRIBAL CONTRACTOR
A	1,500 cu. m	557 cu. m
B	0 cu. ft.	845 cu. m
C	1,800 sq m	5878 cu. m
D	700 sq m	700 sq m

Klallam Tribe began the process of developing and signing the legal documents required by the NHPA: a memorandum of agreement that would lay out the fundamental parts of the agreement and a treatment plan that would identify the specifics of the archaeological work (how much excavation, excavation strategy, expertise to be used, and artifact analysis methods).

Developing the agreement documents took longer than anticipated. There were a number of signatories involved, the archaeological situation was complex, and the parties assigned different levels of importance to the resources. Meetings were emotionally charged, which led to more meetings and more and longer document reviews. After several delays, the signatories signed the agreements on March 16, 2004. In all, 10 agreement documents were prepared and signed or reviewed for the project (Figure 2.6).

The memorandum of agreement acknowledged the archaeological importance of the site and the cultural importance of the site to the tribe. The memorandum of agreement also committed to mitigate the effects that the project had on the cultural deposits; these details would be included in the treatment plan. The treatment plan of 87 pages identified $4.6 million worth of archaeological excavation to be conducted during a 14-week field effort and a 24-month analysis and report phase. The goal of the treatment plan, or archaeological recovery plan, was to answer questions concerning the age of the village, the length of occupation, the activities that took place there, and the spatial extent of the archaeological deposits. It also committed funds to appropriately remove the archaeological remains of the village important to the tribe.

A separate agreement between WSDOT and the Lower Elwha Klallam, called the mitigation agreement, was also signed on March 16, 2004. It stipulated that WSDOT would pay the tribe $3,437,000 in reburial mitigation funds and write letters supporting requests to the federal government for additional mitigation funds. In return, the Lower Elwha Klallam agreed to a legal release discharging WSDOT from any future claims and a covenant not to sue.

With the agreements in place, planning for the work commenced immediately. Graving dock construction would continue in parts of the site where archaeological remains did not appear to be affected. The construction contractor started mobilizing soon after the agreements were signed. The archaeological contractor's plans called for its work to be conducted between April 16 and mid-August 2004. The expectations were that the archaeological remains would be contained in Areas A, B, C, and D; that the only human remains discovered would be fragments from the 1915 mill construction; and that archaeological finds would not slow future construction.

New Archaeological Scope

While the organizations were negotiating the memorandum of agreement and treatment plan, a new project surfaced associated with the graving dock project. As the rainy season arrived, the project area became wet and swampy. By November 2003, it became clear that a drainage system was needed. The concept was to construct

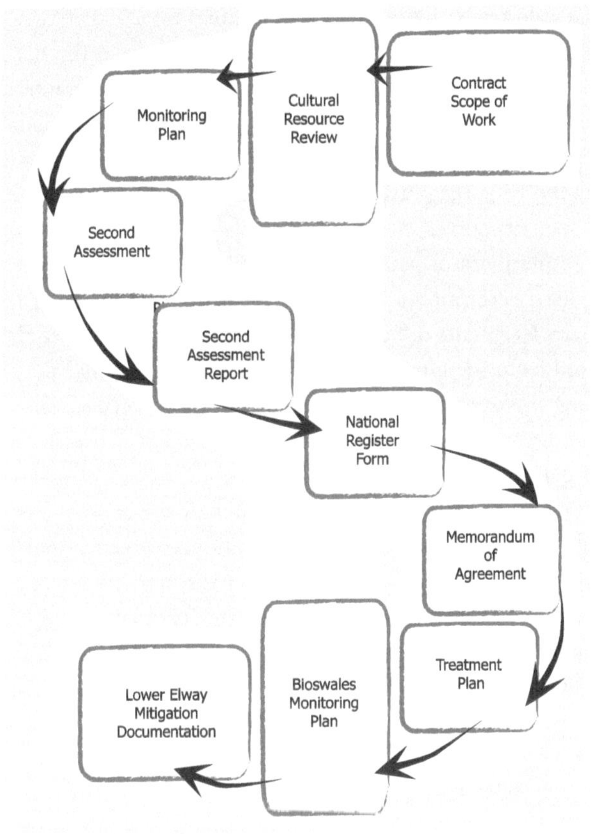

FIGURE 2.6. *The Port Angeles Graving Dock project required numerous plans, reports, and agreements, which collectively contributed significantly to schedule delays and cost overruns.*

large bioswales, essentially an engineered ditch filled with gravel and vegetation to slow, channel, and filter water. The location of the drainage project was outside the originally defined project area, and thus, no archaeological characterization work had been conducted there.

The WSDOT, the Washington SHPO, and the Lower Elwha Klallam developed a separate plan whereby work on the drainage project would begin, with archaeologists monitoring the construction. Protocols used to monitor the construction would be evaluated and adapted for use when the main construction began. Work on the drainage and bioswale construction began November 11, 2003. Wet conditions and high groundwater put the six-week project behind almost from the start.

Just as the archaeologists were preparing to start further characterizing areas A, B, C, and D, the archaeological monitors supporting the drainage and bioswale project began unearthing human remains. On April 19, 2004, the first undisturbed, intact burial found during the graving dock project was discovered. During the next two weeks of construction, two dozen burials were located and removed. Meanwhile, large numbers of artifacts were being recovered in the main project area. Archaeologists were finding fire hearths, storage pits, cooking areas, midden deposits, and many fragments of human remains. By mid-May, more than 50 burials had been removed from the bioswale area. The team of 35 archaeologists and tribal members working in the main project area continued to recover large numbers of artifacts from undisturbed prehistoric midden (prehistoric trash) deposits. Costs escalated out of control.

The increasing number of discoveries heightened the concerns of the tribe. In late May, the tribe formally requested to expand efforts to recover burials within the project area, even in those areas not slated for construction excavation. For the next two months, archaeologists continued to find prehistoric village deposits and burials. By mid-August, when the archaeological work was supposed to be completed, 160 fully intact burials and 440 partial remains had been recovered. The archaeological schedule was extended to mid-September; however, work continued into November 2004.

In late August 2004, the tribe wrote WSDOT requesting that construction work pause until all of the burials on the site were recovered. By that time, 256 intact burials and 442 isolated discoveries of skeletal material had been recovered. WSDOT did curtail some construction activities, but other work continued into October, as the tribe and WSDOT negotiated the scope of additional characterization work to be conducted at the site. But as discoveries of remains continued, before long it became obvious to the Lower Elwha Klallam Tribe that the project had to stop. The tribe exercised the formal dispute resolution clause of the memorandum of agreement.

Project Shutdown

Discussions, letters, legal opinions, and public frustration continued through November. On December 10, 2004, the tribe formally requested that the project at the Port Angeles Graving Dock be terminated. On December 21, WSDOT agreed to terminate the project and pursue another location for the graving dock. Approximately two years after work on the graving yard project had started, the project ended in failure, with considerable cost to all involved.

The archaeological profession had not fared well. The archaeologists and the oversight system had not provided adequate input to the project. The techniques and characterization sampling strategy developed for the 22.5-acre parcel failed to identify one of the largest prehistoric villages ever encountered in the Pacific Northwest. Thousands of artifacts from the village site and the remains from more than 500 tribal ancestors had been disturbed. Costs of the archaeology and repatriation of human remains exceeded $8 million. The delays, numbering over 400 days, led to over $60 million in additional construction project costs. The delays to the Hood Canal Bridge, deemed to be "beyond its engineered life," posed additional risks to the drivers using the bridge for more than a year.

Everyone was upset. The tribe was upset at the desecration of its ancestors and ancient village. The local community was upset about losing a $19 million project that would provide jobs and general economic relief to the community. Everyone was upset at WSDOT. The WSDOT secretary eventually resigned, due in no small way to the graving dock expenditures. The local press provided regular coverage of the project, while the Seattle papers published stories and editorials at major points in the process (Table 2.2).

Because of the extent of problems surrounding the graving dock case, the state of Washington commissioned two formal reviews of the project and the mistakes that had been made, each of which resulted in recommendations to the various state agencies involved. WSDOT conducted one study (WSDOT 2006), and the Washington State Joint Legislative Audit and Review Committee commissioned the other (JLARC 2006).

Over the ensuing months, the state worked with the tribe to repatriate the human remains. The state ultimately transferred the 22 acres to the tribe for use as a tribal cultural center and monument to the people who had been buried there.

Observations

Here we pose several observations about the Port Angeles Graving Dock project to set the stage for later discussions on archaeological risk management.

- The first request to conduct background research and a field survey, make contact with the tribe, and prepare a report within a three-week period was unrealistic and set an unnecessary tone of urgency.

TABLE 2.2. Selected newspaper headlines from the Port Angeles Graving Dock project showing the escalation of project costs as cultural remains were encountered

DATE	NEWSPAPER HEADLINE	PROJECT ESTIMATED COST*
11-20-2002	Port Angeles: Area elated as state selects Port for graving yard facility	$30 M/ $19 M
06-24-2003	$200 million-plus Hood Canal Bridge/graving yard contract could be awarded this week	$204M/ $19 M
08-03-2003	PA graving yard groundbreaking set Wednesday	$204 M/ $19 M
08-29-2003	Discovery of human remains halts bridge project (ST)	$204 M/ $17 M
09-01-2003	Port Angeles: Burning of feast table resurrects forgotten tribal ceremony	$204 M/ $17 M
09-02-2003	Port Angeles: Digging of 200,000 cubic foot graving yard awaits archeological plan	$204 M/ $17 M
09-05-2003	Few alternatives for Port Angeles graving yard; other locales don't meet giant needs	$204 M/ $17 M
09-12-2003	Tribe, state reach pact on graving site	$204 M/ $17 M
09-24-2003	Port Angeles: Archeological survey finding Native American items at graving yard site	$204 M/ $17 M
10-26-2003	When can the graving yard construction resume?	$204 M/ $17 M
11-06-2003	Sides in graving yard negotiations meet again; little progress reported	$204 M/ $17 M
11-07-2003	Tribal remains used as backfill at mill in 1915	$204 M/ $17 M
11-09-2003	Breakthrough on graving yard talks: Tribe, state OK drainage work, partially ending 2½-month shutdown	$204 M/ $17 M
12-16-2003	Graving yard talks in fourth month	$204 M/ $17 M
03-16-2004	Tribe to begin search for remains at construction site (SPI)	$204 M/ $17 M

*The first number is the estimated total cost of the bridge replacement project; the second is the amount that the Port Angeles Graving Dock portion of the project was estimated to be. Both numbers got revised as costs escalated due to the archaeology and burials.

DATE	NEWSPAPER HEADLINE	PROJECT ESTIMATED COST*
03-18-2004	Area leaders pleased, relieved that graving yard back on track	$271 M/ $17 M
03-23-2004	State buys graving yard site from Port for $4.84 million	$271 M/ $17 M
04-11-2004	Tribal chairman outlines how $3.4 million in 'mitigation' funds will be spent	$271 M/ $17 M
04-20-2004	Archaeological workers begin poring over graving yard site	$271 M/ $17 M
05-14-2004	Remains of more than 50 Klallam ancestors uncovered at graving yard site	$271 M/ $17 M
07-14-2004	Dig site unearths ancient Klallam village (SPI)	$271 M/ $17 M
09-26-2004	Graving yard archaeology to continue into at least November	$272 M/ $17 M
10-08-2004	Lower Elwha want temporary construction shutdown at graving yard	$272 M/ $17 M
11-14-2004	Graving yard, archeological dig mired in delay while federal decision awaited	$272 M/ $17 M
12-12-2004	Lower Elwha Klallam tribe asks Department of Transportation to move graving yard out of Port Angeles	$272 M/ $17 M
12-15-2004	Tribe wants all digging, including archaeological, to end at graving yard site	$272 M/ $17 M
12-22-2004	State drops graving yard project	$272 M/ $59 M
01-04-2005	State transportation officials, tribe to meet today over future of Port Angeles graving yard site	$272 M/ $59 M
01-07-2005	State lawmakers urge probe into $58 million graving yard abandonment	$272 M/ $59 M
01-14-2005	Lower Elwah to host healing ceremony at former graving yard site on Saturday	$272 M/ $59 M
01-16-2005	Tse-whit-zen existed before Christ walked the Earth	$272 M/ $59 M

Continued on next page

TABLE 2.2. *Continued*

DATE	NEWSPAPER HEADLINE	PROJECT ESTIMATED COST*
01-20-2005	No chance to restart Port Angeles graving yard, state Transportation Commission told	$272 M/ $59 M
01-28-2005	Gregoire supports investigation of $59 million graving yard mistake	$272 M/ $59 M
02-25-2005	Governor to Mayor: Graving yard isn't going to resume at Port Angeles site	$291 M/ $59 M
03-06-2005	Early 1990s archaeology reports tell of presence of *Tse-whit-zen*	$291 M/ $59 M
05-22-2005	"Unearthing of Tse-whit-zen" tells an extraordinary story (ST)	$291 M/ $59 M
07-14-2005	Firm to visit Port Angeles to investigate graving yard situation	$291 M/ $59 M
08-14-2005	Lower Elwha Klallam sue to have graving yard site remain as burial ground	$291 M/ $59 M
11-30-2005	Lower Elwha, Department of Transportation take different strategies toward graving yard resolution	$291 M/ $59 M
12-23-2005	State, Lower Elwha agree to negotiations over graving yard site	$291 M/ $59 M
02-05-2006	Hood Canal Bridge price tag jumps 62 percent to $471 million	$471 M/ $59 M
08-14-2006	State, tribe settle lawsuit over Port Angles Indian graveyard (ST)	$291 M/ $59 M
08-15-2006	Graving yard deal: $15 million	$291 M/ $85 M
08-15-2006	Lower Elwha get land to rebury their ancestors (ST)	$291 M/ $85 M
04-28-2007	State highway chief resigns	$291 M/ $85 M
09-16-2008	Lower Elwha tribe reburying 300 ancestors at village site in Port Angeles	$291 M/ $85 M

- The budget for the initial assessment ($6,700) was too low to complete a full NHPA Section 106 review. Considering the size of the project (millions of dollars and 22.5 acres), the sensitive location (on the waterfront in an area historically and presently occupied by Northwest Coast Indian people), and the consequences that might result from the discovery of important archaeological deposits (construction delays and cost escalation), the one-week effort to characterize the project location could only be viewed as a preliminary assessment, and designed to identify the work that would still need to be done to complete a good assessment for the project.
- The recommendation to move ahead with the project without additional archaeological characterization and to rely on archaeological monitoring of construction below certain depths was premature. The amount of subsurface sampling done to determine if archaeological deposits were located within the area of potential effects was surprisingly small. The three-day sampling effort was inadequate to predict the likelihood of discovering archaeological resources once construction began. Archaeological monitoring should be recommended only when the risk of discovery is low. Nevertheless, the recommendation to monitor construction activities and stop work if significant remains were encountered was made, and the SHPO and tribe concurred.
- Basic important information about the project site—specifically, the location of the historic beach where cemeteries were located—should have been generated and would have greatly improved the quality of the initial two phases of archaeological site characterization efforts (Figure 2.7). The failure of the two characterization efforts to locate the largest Native American cemetery ever found on the Northwest Coast was pivotal and speaks volumes about archaeological methods and compliance requirements.
- Contrary to common professional practice, the defined area of potential effect restricted archaeological characterization efforts to a 9-acre area, even though it was obvious there would be ground disturbance across the entire 22-acre parcel. The ancillary facilities were excluded from the original project descriptions because their specific locations were still to be determined, yet they were clearly going to be situated somewhere within the 22.5 acres. In fact, the initial site clearing caused considerable impact across the project area. In retrospect, the entire parcel should have been evaluated for impacts to archaeological resources, not just the 9-acre area.
- Agreement documents for projects involving numerous parties took a long time to prepare and still failed to provide the direction needed. The assumptions made about the horizontal and vertical extent of archaeological resources—the implicit

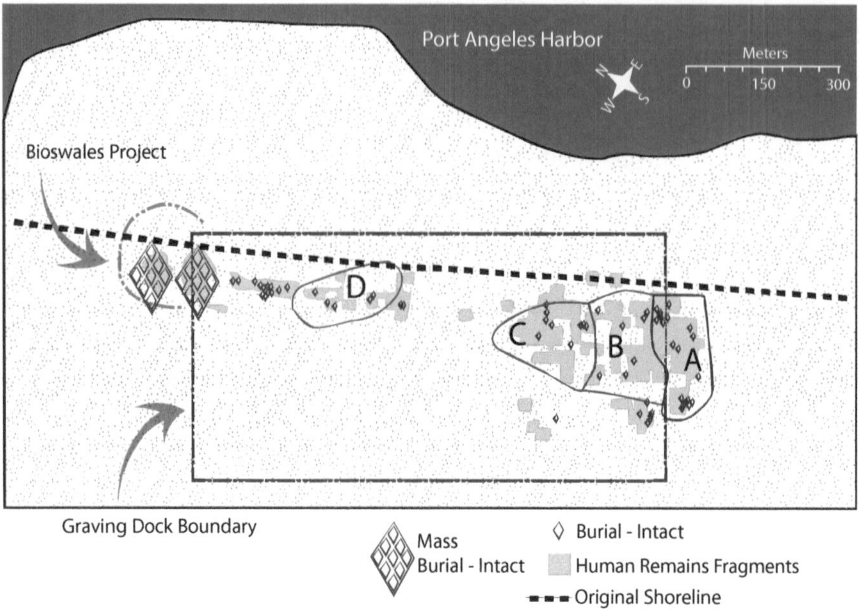

FIGURE 2.7. *Map of the Graving Dock project area showing original shoreline and proximity of archaeological and burial areas to that shoreline. Had the location of the original shoreline been identified during the initial stages of the project, a different sampling strategy might have been deployed and the locations of many of the burials discovered prior to the start of construction.*

conceptual models of the archaeologists—were quickly disproved as subsurface information began to emerge, yet no adjustments were made. In particular, the parties failed to predict the presence of human remains, despite substantial evidence of this possibility.

- No one ever seemed to seriously address the question, "Have we adequately characterized the project area for cultural resources?" If they had, especially given the potential consequences that a discovery of cultural resources could have on such a project, it is likely that additional archaeological sampling would have been authorized (Figure 2.8).
- Despite the widespread outrage of tribes and archaeologists following the Blaine events of two years earlier, history repeated itself within two years at Port Angeles. Several lessons learned from the Blaine project—including the need for meaningful tribal consultation, the need to adequately sample and characterize a site, and the need for appropriate archaeological expertise—went unheeded. If the failures at Blaine had been communicated to others involved with projects

Seattle Post-Intelligencer

http://seattlepi.nwsource.com/opinion/203575_gravesited.html

Avoid doing wrong again

Tuesday, December 14, 2004

SEATTLE POST-INTELLIGENCER EDITORIAL BOARD

As important as it was to redress the historical wrong done in the Chief Leschi case, it is also important to avoid yet another historical wrong at the village of Tse-whit-zen at Port Angeles.

The Lower Elwha Klallam Tribe is right to ask the state Department of Transportation to end excavation and construction work at the proposed graving dock there. The Hood Canal Bridge repair project is important and the investment lost at the Port Angeles site is massive. But the human and archaeological issues overshadow both.

Gov. Gary Locke said, "Had everyone known the magnitude of the archaeological significance when this was first proposed and when the testing for the site was done, nobody would have gone forward with it."

Well, now we know. And the work at that site should not go forward.

The governor's remarks raise pertinent questions about the archaeological survey work going into the project. The sad bounty of human remains and other artifacts uncovered during the initial construction made it clear that this was a potentially rich find. The state should evaluate thoroughly a process that apparently managed to miss a dramatic archaeological discovery in state history.

The state could fight the inevitable, but there would be no real victory in such a fight, only losses, of perhaps historic proportions.

FIGURE 2.8. *Editorial from the Seattle Post-Intelligencer Editorial Board calling for an evaluation to understand how archaeological deposits and burials could have been missed prior to the start of construction.*

located in culturally sensitive areas in the region, the disaster at Port Angeles might not have occurred.

SUMMARY

In this chapter we have presented detailed descriptions of two high-profile cases from the Pacific Northwest of North America. We have used the Port Angeles Graving Dock and Blaine Wastewater Treatment Facility Expansion projects to illustrate the types of problems that can arise when development projects are located in archaeologically and culturally sensitive areas. In the following chapters, we look at the stages of a typical construction project and explore the types of actions that project managers can take to minimize the potential for archaeological problems as their projects evolve.

Note

1 The information provided here is based on the investigation conducted by the Register of Professional Archaeologists, following accusations that the lead archaeologists had violated the code of ethics (McGimsey, Eisenberg, and Doershuk 2003).

3 Addressing Archaeological Risks during Early Project Planning

When we review projects that have encountered major archaeological problems, it is usually obvious that the problems could have been avoided if more archaeological work had been conducted before construction began. In most cases there were plenty of indications that archaeological resources might be present. Either the project was located in a cultural area with a high potential for resources, or a warning had been issued by an interested person. But, for whatever reason, project personnel did not take the appropriate measures to assess the project location and determine whether important resources existed.

As soon as project personnel learn that sensitive resources may be impacted by project activities, the project manager needs to assign a point person on the team to be responsible for archaeology and heritage resources. This person will have to coordinate a number of activities, most of which will be associated with an initial cultural assessment. It is to the project's advantage to have a staff member who has experience with archaeological and other heritage resource compliance issues, but in reality, few companies or organizations have such experts available. More typically, responsibility is assigned to someone with little or no experience with such issues. This makes it all the more important to obtain a good initial assessment of the potential project impacts, an assessment produced by a qualified heritage resource professional and one that addresses the likelihood of resources being present, the consequences of a discovery, and the regulatory compliance path forward.

When it appears that archaeological or heritage resource work must be done in support of the project, the manager needs to develop a budget for it. There will be meetings to attend, letters to write, reports to be written, and contracts to let and monitor. Initially, costs may be small enough that they can be borne by the project management or regulatory group; at some point, however, both a separate line item in the overall project budget and a separate, detailed budget should be created for the archaeological work. Because of the uncertainty involved in projects located in areas with archaeological deposits, it is important to include contingencies in the budgeting process.

In this chapter, we focus on four activities that the project manager and the person assigned to oversee cultural resources will need to perform if a project is located in a culturally sensitive area: obtain expertise, initiate contact with others, define requirements, and produce the assessment.

OBTAINING EXPERTISE

Hiring archaeological or other heritage resource expertise as soon as you suspect that your project will have those kinds of issues is an important first step when you have no expert on staff. Depending on where the project is located, this may be easy or difficult. In many countries, all archaeological work is done by government archaeologists. In some countries, archaeological work has been privatized, and many archaeological contracting firms—ranging from one-person companies to firms employing dozens of archaeologists, historians, architectural historians, graphic artists, editors, and administrative staff—are available to do what needs to be done. Large environmental and engineering firms are increasingly adding archaeologists to their staff for in-house, and occasionally outside, projects. In some countries, universities may be the best place to find archaeological expertise. Searching the Internet and making a few calls should get you on the right path.

Professional Qualifications

Wherever you find your archaeological experts, it is important to make sure they are well qualified for the job at hand. Minimal qualifications include:

- *A technical degree* in heritage resource management, anthropology, archaeology, or a related discipline. If archaeological issues are the focus, you need someone with a background in field archaeology. If the work is located in places that have important architectural resources, an architectural historian will be needed. If the project location contains ethnohistoric resources important to a local indigenous group or other group with historical ties to the area, an ethnohistorian or cultural anthropologist is needed. If you might be dealing with several types of resources, however, you will need access to a broad base of expertise.
- *Experience working in a project-focused setting.* The professional you hire needs to be well versed in the regulatory environment of your project area. Your focus is not research per se; you are trying to comply with a variety of requirements and make sure the project activities do no harm. The expert needs to be comfortable with all aspects of compliance work and should be able to work in the risk management mind-set. An academic archaeologist can likely advise you on the tech-

nical merits of the project area or affected resources but may have trouble working within the compliance system.

- *Standing as a professional.* Various countries and professions have specific qualifications that need to be met before a person can claim to be a "professional." Some countries have organizations or lists of members who have met their requirements, which are often based on education and experience. The Register of Professional Archaeologists in the United States is one example. Organizations that have been formed around the world are found in Appendix B. Project personnel should recognize that professionals have ethical standards that they must abide by in performing their work; we have included a selection of standards as examples and provided them in Appendix C. Ethical standards such as these are critical if the profession is to maintain its independence and avoid the perception that its practitioners can be bought (see, for example, the recent claims by King [2009] that the American cultural recent compliance industry is corrupt).
- *Local expertise.* Archaeology and many other heritage resource disciplines are often location-specific. Someone from Washington state should not be spearheading work in Hawaii unless he or she has worked there before. The professional doing the assessment must have experience in that area in order to conduct a professional job identifying and evaluating the resources. It is not always possible to find someone with local expertise, in which case regional expertise is a must. Once again, a project manager needs to be aware enough to know when local expertise should be added to the team.
- *Access to supplemental expertise.* Heritage management has grown into a field composed of a number of very technical disciplines, each with highly specialized subdisciplines. In archaeology, one may need experts in excavation, artifact analysis, geomorphology, soils, and it goes on from there. It is not always possible to have staff with the full range of expertise, so it is important to find professionals who have access to specialists on call.

Contracting Approaches

Once the archaeological expertise has been found, the project manager must get the heritage firm under contract. Contracting for heritage expertise is similar to contracting for other services. Depending on the organizational structure, contracting for resources may be simple or complex. If government agencies are involved, it will likely be complex; various contracting rules will have to be followed (in the United States, they are known as the *federal acquisition requirements*, or more commonly, the FARs). If the work has to be bid competitively, a proposal package will be prepared, firms will bid on the work, and a selection will need to be made.

Alternatively, sometimes a sole-source (or no-bid) contract can be awarded if certain criteria are met—for instance, if the dollar value of the contract is below a certain threshold, or the company being awarded the contract is a small business or minority owned. Private companies (as opposed to governmental agencies) typically have more flexibility in their contracting mechanisms and can contract directly with a firm if they find one that meets their needs.

For archaeological situations, a project manager should pay particular attention to the following principles:

- use the appropriate contracting mechanism;
- write well-defined scopes of work;
- base contract awards on multiple factors.

If these principles are followed, there will be minimal complications, and the process of contracting for archaeological expertise will stay on schedule.

Using the Appropriate Contracting Mechanism

Agencies and companies use a variety of contracting approaches to obtain services. Some contracts are one-time-only, single-project arrangements, while others run throughout some period of performance (e.g., two years, five years) and may cover multiple projects. Where long-term contracts have been established to provide support for different projects, individual statements of work will still need to be prepared, and the appropriate level of effort, number of hours, miscellaneous costs, and such (often referred to as *task order agreements*) will have to be negotiated. This was the case at Port Angeles, where the first archaeological contractor had an "on call" agreement; the agency had gone through a contracting process years earlier to prequalify different heritage resource contractors for future work with the agency. With rates, insurance, staff qualifications, and other details already agreed upon, all that remained to be settled was the scope, approach, and level of effort for the task at hand.

Two common types of contracts used for archaeological services are *fixed price* and *cost reimbursement for time and materials*. In fixed-price contracts, the project contracting officer issues a request for proposal (RFP), which includes the information contractors need to prepare a bid, such as the schedule, scope of work, location, and any special requirements. Contractor(s) reply with a description of the approach they intend to follow to meet the scope, and a price for which they will do the work. With fixed-price contracts, the contractor takes on the risk; if the archaeological firm winds up being unable to do the work for the agreed-upon price, the archaeological firm must bear the additional cost. On the other hand, if it can complete the work for less than the specified price, it makes more profit.

For archaeological projects, the preferred type of contract depends on the type of services required and the amount of uncertainty embedded in those services. For example, if an initial archaeological assessment is needed for the project, and the contractor is able to give a reliable estimate for the cost of work that needs to be done, then a fixed-price contract is ideal for both parties. The contractor can specify the number of hours of background research needed, the number of meetings, the acreage to be surface surveyed and shovel tested, and the time to write the report. In this case, the contractor can be reasonably certain how long it will take to accomplish the work, and can add an additional line item for contingencies to cover unexpected costs, such as delays from bad weather. The contractor submits the cost proposal, the client accepts it, the contractor writes the assessment, and the client pays the contractor. The advantage to the project is that costs are fixed.

For proposed archaeological work where there are more uncertain factors, cost reimbursement for time and materials may be a better option. For example, if the requested services involve removal of an archaeological deposit for which the extent is not fully known, it is nearly impossible to predict with any accuracy how long data recovery will take. A prudent bidder will be conservative, allowing for the possibility that the archaeological deposits will be extensive, meaning the bid will be high. With the cost-reimbursement contract, if it turns out that few archaeological remains are found, the contractor would be paid only for the work actually undertaken.

Generally, if your project team has someone with experience in heritage resource compliance who can work with the contracting officer to oversee the archaeological contract, and you have an archaeological contractor who you know does good work, a cost plus fixed-fee contract could be the least costly, most profitable, and least risky option for the project.

Writing Well-Defined Scopes of Work

The scope of work is the defining document identifying the work that needs to be accomplished. The more specific the scope of work, the easier it is to determine whether or not that work was completed. This may sound simple, but it is not always an easy matter to define the details for archaeological work. In the Port Angeles initial assessment, for instance, the agency and contractor took over three weeks to agree on the scope of work.

When we review archaeological projects that have run into problems, we often find poorly defined scopes of work. It is not uncommon for clients and regulators to complain that the work accomplished was not what was expected; and when the scopes of work are reviewed retrospectively, it is clear that the scopes were in fact too general. To write a well-defined scope of work, someone on the project has to have enough experience with archaeological compliance to know what needs to be done.

If no such person is available on staff, a call to the regulator can help define the work that is needed.

Some of the common details found in good Requests for Proposals (RFPs) are:

- clear direction on the area(s) to be investigated (sometimes referred to as the *area of potential effect*), including maps to reduce confusion;
- range and level of background research that will be done;
- expectations concerning interviews or meetings with outside parties;
- expectations concerning the range and level of expertise to be used;
- expectations concerning subsurface sampling strategy to be used;
- expectations concerning documentation and final report.

To the degree that scopes of work can be specific and tailored to address archaeological risks facing the project, the better off the project will be.

Basing Contract Awards on Multiple Factors

In situations where the project contracting officer requests bids from more than one archaeological contracting firm, there must be a set of criteria on which to base the award. Typical criteria include ability to do the work, qualified staff, past performance, and cost. In many cases, it comes down to cost, and usually the low bidder wins. The lowest bid is not necessarily a poor choice, if the scope of work is detailed and clear and the winning bidder's qualifications meet the requirements.

It is important that archaeological RFPs specify any critical requirements in their award evaluation criteria. As discussed earlier, a special expertise may be vital to a project. In the Blaine Wastewater Treatment Facility Expansion project, there was no well-defined requirement to hire someone experienced with human remains. For a project that would be working in and around a known American Indian cemetery, that omission was risky. At Port Angeles, there was no well-defined requirement for someone with geomorphology experience; this may be why no one on the project team thought it important to identify the location of the historic shoreline.

Another important criterion is experience working with heritage resources in the project area. At Blaine, for the second phase of work, the contractor team with little experience in Northwest Coast archaeology won out over the local contractor who was well versed in the local and regional archaeology—only because it had a lower bid. Had the original firm won the second phase of archaeological work at Blaine, the project may not have deteriorated the way it did.

The project manager must also consider a bidder's stated approach for meeting the scope of work. Is the proposed approach adequate to meet the scope? For example, if the scope includes identifying archaeological deposits under a proposed football stadium, and one bidder says it can be done with 500 shovels tests, and another bidder says it will take 1,000 shovel tests, which one is closer to the number needed for a reliable subsurface sample? Factors such as adequacy of the methods proposed need to be factored into the contract evaluation and award process.

INITIATING CONTACT WITH STAKEHOLDERS

Groups and individuals who may have concerns about impacts to heritage resources need to be contacted early in the process. While this task could be deferred to a contractor, it is better for the project manager or designated team member to initiate contacts. Some stakeholders, such as regulators, may be easy to contact. Others, such as descendants, indigenous groups, religious groups, or local preservationists, may be less organized and less accessible; it may take a long time to identify the right people to talk to and may require considerable effort to solicit their concerns, interests, and expectations.

Sometimes the responsibility for official contact, or consultation, is defined by law and may dictate that certain organizations and levels of management be involved. In the United States, for example, federal agencies are required to consult with the state historic preservation officer, Indian tribes, and Native Hawaiians and Alaskans. Consultation may need to be "government to government." In these cases, elected officials should meet with elected officials, policy people should meet with policy people, and technical staff should meet with technical staff. This does not mean others involved in the project—such as the company responsible for doing the construction—can't communicate with the various players. That communication should and often must occur; but it is important to recognize that such communication needs to happen within the formal consultation process. You may find it useful to think of formal consultation as "Consultation with a capital C," and informal consultation as "consultation with a small c." Regardless of the approach taken, experience shows that communication with others about archaeological and cultural sensitivity is probably the single most important effort that a project manager can make to prevent major archaeological problems.

By the time a project officially begins, there may already be indications that the project is located in a culturally sensitive area. In these cases, the project manager will likely already have identified some of the stakeholders with whom it is important to discuss the issues, concerns, and expectations. Based on this input, the project team can get a sense for how soon certain things may need to occur, the individuals or organizations that should be contacted, and the level of effort that might have to be made to protect important resources.

In other cases, a project manager may have no indication about cultural sensitivity of the project area at the start of the project. This should not be interpreted as an indication that the area is not sensitive, but rather that an effort is needed to get some guidance from local officials or governmental historic preservation authorities. Based on the input received, the project manager can make an initial judgment about the nature and urgency of future efforts.

IDENTIFYING REGULATORY AND OTHER REQUIREMENTS

Many nations and regions have laws to protect heritage resources and require projects to get some kind of "green light" approval once it is demonstrated that important resources will not be destroyed by project activities. Some countries may require an actual permit before work can begin, while other places may only require notification or consultation. Some countries may require that a state agency review the project, while other countries may not have any governmental office with which to consult. Appendix D provides information about heritage resource requirements that we found for countries around the world.

Within nations, there can also be certain groups who have state-recognized rights to lands and resources. A project involving First Nations in Canada, Native Americans in the United States, or Australian Aborigines may be required to consult with the group before and during the project, as explained above. Project personnel should recognize that many groups, especially indigenous groups, also have their own natural laws that guide their interests, concerns, and expectations. Balancing the needs of the various groups with the needs of the project proponents can be challenging.

You will likely need a professional heritage resource person to identify the relevant and applicable regulations and consider how they will affect your project. Identify all of these requirements early in the process so that subsequent actions are built on a solid requirements-defined foundation. Where there are different sets of expectations that need to be met, patience and a tolerance for ambiguity are needed.

Regulatory reviews typically involve three steps:

- First, you need to determine if there are any archaeological or other resources in the project area.
- Second, if there are resources, you have to determine if they are important to anyone.
- Third, if there are important resources, you need to figure out how to avoid them or mitigate the potential impacts from project activities.

Whether the project needs a permit or some other formal document before work can begin depends on the specific applicable laws. In some places, such as the United States, a permit is not needed, but the project proponent is required to complete a review process and obtain concurrence from the state historic preservation office.

Complying with regulatory requirements can get real complicated real fast when important resources are discovered. For example, upon such a discovery, the regulator and outside parties may want to review plans and findings, requiring additional time to receive and resolve comments. Various professional reports may be needed, requiring time and money to prepare. These additional requirements need to be identified early so that a good archaeological compliance approach can be designed.

Whether or not there are formal requirements, a project manager should consider the wishes and expectations of others. Here it is important to realize that heritage resources are not the enemy. Rather, they are important features of our communities, so if a project manager also wants the project to be an important part of the community, then he or she needs to do the right thing by the heritage resources, if at all possible. Thus, it is critical for a project manager to understand the importance of the place that is being developed so that the project team can avoid inadvertently damaging a resource that is important to the community.

PREPARING THE INITIAL ARCHAEOLOGICAL ASSESSMENT

Once the location is secured and there is reasonable certainty that the project will proceed, it is time to commission an archaeological assessment for the project and project area. The purpose of the initial assessment is to develop predictions about archaeological deposits and other heritage resources that might be important to somebody and that might be impacted by the project. It is critical to identify early on the potential areas where such impacts could occur (Figure 3.1).

Project personnel typically like to define the area of potential effect narrowly, which they believe will reduce archaeological costs. For example, consider the case where a large building is being constructed. The project manager may only want archaeological testing in the area where the main building foundation will be located, because that is where the digging will occur. But what about all the pipelines, electrical trenches, sewer lines, roads, and lay-down yards that will be constructed? All of these can impact archaeological remains. And what about other types of impacts? What if the building is located next to a famous historic area or an area considered sacred by some group? Will there be visual effects? Will the new project impact the historical character of the area?

This was the situation at Port Angeles, where the archaeological work was restricted to the area within the proposed graving dock. It turned out that most of the burials were located at the edge of the proposed graving dock, and they were unearthed only when the project expanded outside the original footprint to install drainage features.

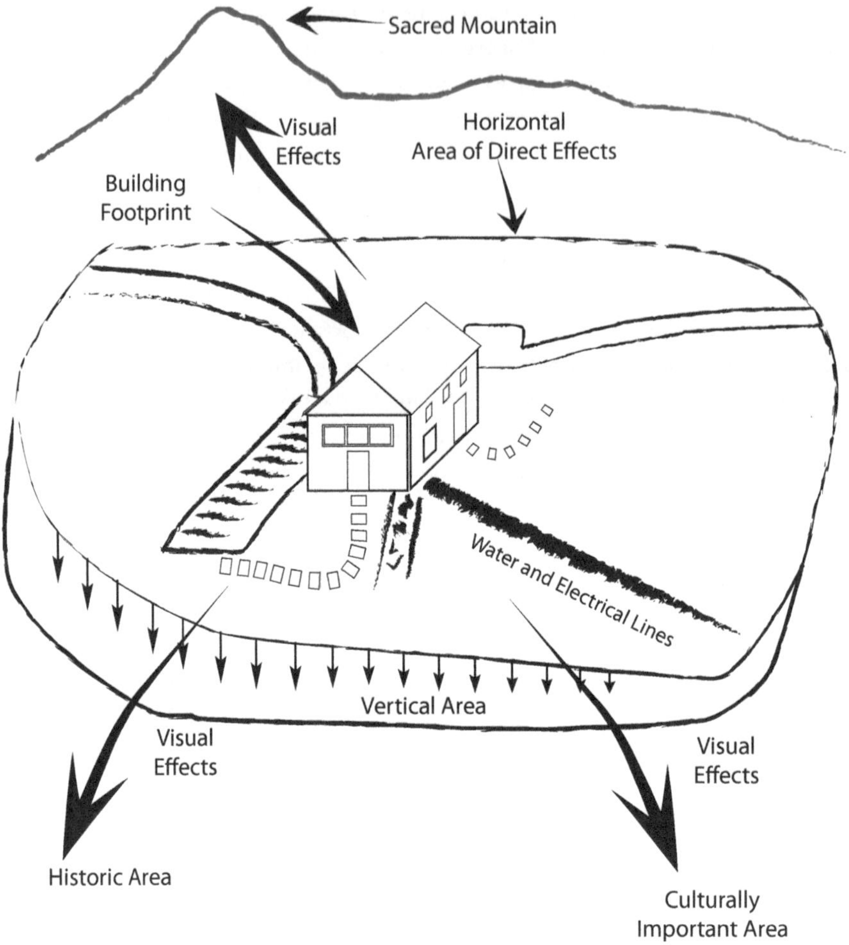

FIGURE 3.1. *Drawing of a hypothetical building construction project, showing that project activities can impact areas and resources far beyond the building footprint. Defining all areas of potential effect is critical to avoiding problems associated with heritage resources once construction begins.*

If questions concerning all types of impacts are not considered early on, the project runs the risk of disturbing resources later, probably during construction. To avoid this, it is best to draw a larger area of potential effect by defining the range of impacts that a project could cause. In some places, the concepts of *direct effects* and *indirect effects* are used. Direct effects are those that occur as a direct result of the undertaking: the impact of construction excavation, the visual impact of a tall building, or the loss of access to an important resource. Indirect effects are those that might occur over time as a result of the project. For example, a new road into a remote area may not directly impact any resource, but as a result of the road being constructed, other archaeological sites in the area may be impacted as the indirect result of the number of people new coming into the area and disturbing the sites.

During the early stages of a project, small budgets may limit the archaeological assessment to existing information about the project location. Nevertheless, at some point during the planning process, an initial assessment report should be prepared that incorporates the following items:

- *Geomorphology model:* What are the landforms located in the project area? How were they created? How old are they? What additional information is needed to sufficiently describe the geomorphology for archaeological purposes?
- *Land-use history:* What occurred at this location in the past? Has there been much earth movement or surface disturbance from past activities? Much of this information comes from documentation that is on file with local governments: old maps, tax records, historic aerial photographs, local histories, real estate records, and local knowledge.
- *Heritage resource conceptual model* (Figure 3.2): Given what is known about the landforms onsite and the distribution of heritage sites in the adjacent areas, where might sites be located, how likely is it that they are located there, and what evidence is needed to confirm or reject the predictions? A trip to the central cultural site file archive, if one exists, should reveal any previous work done in the area and the findings.
- *Ethnographic survey:* A survey of cultural groups will identify important places in the project area and collect information from local community members concerning their importance.
- *Project area reconnaissance:* An on-the-ground inspection is usually money well spent so that the archaeologist can get a feel for the area and inspect the surface. A visit to the project location also enables the archaeologist to collect logistical information that will be useful in designing field characterization efforts.
- *Risk assessment:* What is the likelihood that important resources will be impacted by the project, and what will be the consequences if such resources are discovered during the course of the project?
- *Recommendations:* What are the appropriate next steps?

With the initial assessment report in hand, the project manager can now incorporate archaeological risks into the overall project design. The project manager, project engineers, and archaeological contractors can work to locate parts of the project in less

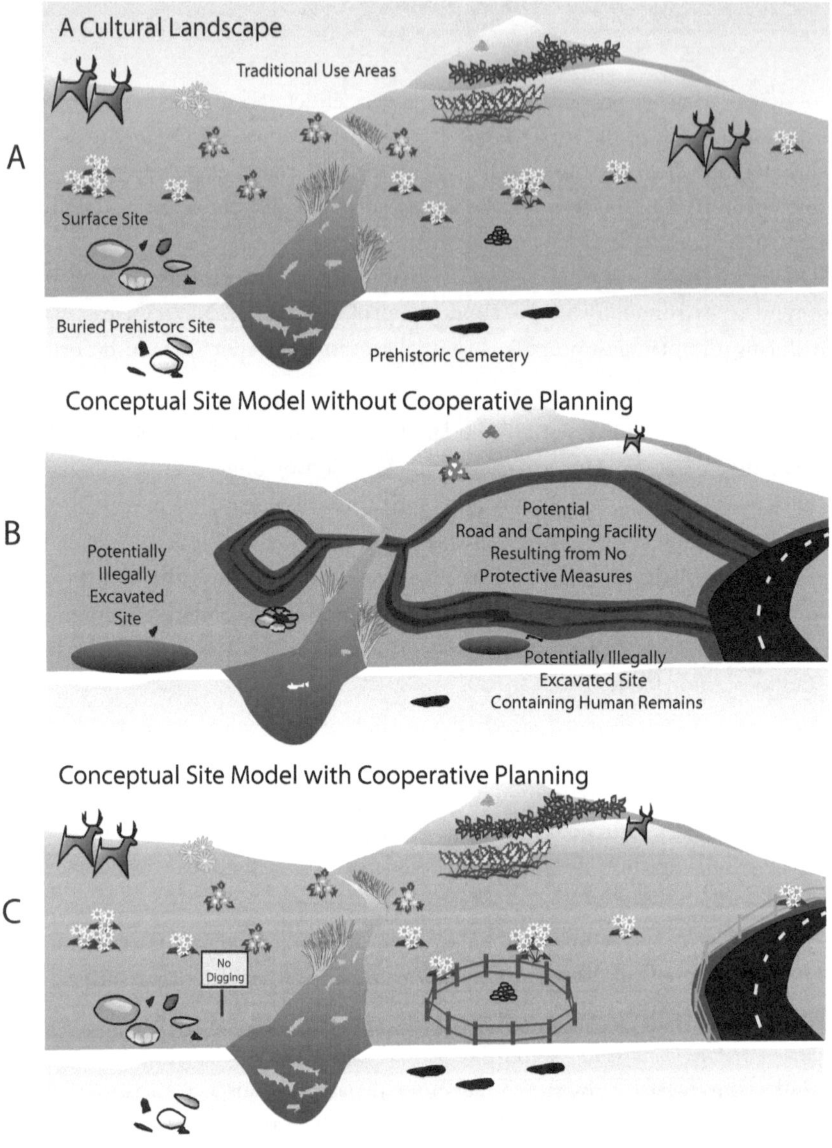

FIGURE 3.2. *An initial archaeological risk assessment produces conceptual maps or models of heritage resources within the proposed project area. It is critical that the maps, which are used to guide future sampling efforts and project design, reflect cooperative work among specialists and interested groups to help all parties to understand the resources that could be in the area. (A) Depiction of a cultural landscape showing the resources located in the area. (B) A conceptual map of the project area made by project personnel without cooperative planning. (C) A conceptual map of the project area made by project personnel with cooperative planning, showing modifications to project design to avoid impacts to important resources.*

sensitive areas, or modify certain design elements to reduce the visual impacts of the project elements. The assessment is also useful for helping design any archaeological sampling efforts that will need to occur to improve the understanding of what is there.

SUMMARY

In this chapter, we have looked at the actions a project manager should take at the start of a project. Early on, a project manager must determine if the project is located in a culturally sensitive area. The project manager must obtain the appropriate expertise so that an initial heritage resource assessment report can be prepared. It is also important for the project manager to communicate with regulators and interested groups to get an early indication of sensitivity and appropriate responses. Finally, after completing the archaeological assessment, the project manager can identify actions that need to be taken to reduce risks and integrate the findings into the general project plan.

4 Implementing Archaeological Measures to Reduce Risks

This chapter is, in many ways, the most important one in the book. In the following pages we go to the heart of the matter: determining if archaeological deposits are located within the project area and in danger of being disturbed. Project managers typically are not involved in designing the archaeological characterization strategy; but we argue that project managers *should* be involved and engaged enough to understand the sampling strategy used to characterize the project area and interpret the results. We say this because it is usually the project manager who best understands the ground-disturbing activities that are planned.

In the following pages we examine several basic archaeological concepts to help project managers decide what archaeological work needs to be conducted, and how best to accomplish that work (Figure 4.1).

The chapter is organized as follows:

- *Identifying Resources in the Project Area:* Approaches that can be taken to locate important resources.
- *Mitigating Effects:* The types of mitigation efforts that might be needed to address impacts to the resources and move the project forward.
- *Archaeological Monitoring during Construction:* Factors considered when deciding whether or not to stop collecting archaeological data and rely on monitoring during construction, risking a stop work order if archaeological deposits are encountered. If archaeological deposits are encountered during monitoring, work may need to stop
- *Addressing Inadvertent Discoveries:* Procedures that need to be followed if unanticipated archaeological deposits or human remain are encountered during construction.
- *Disseminating Findings.* Information needs to be documented and made available to heritage resources databases.

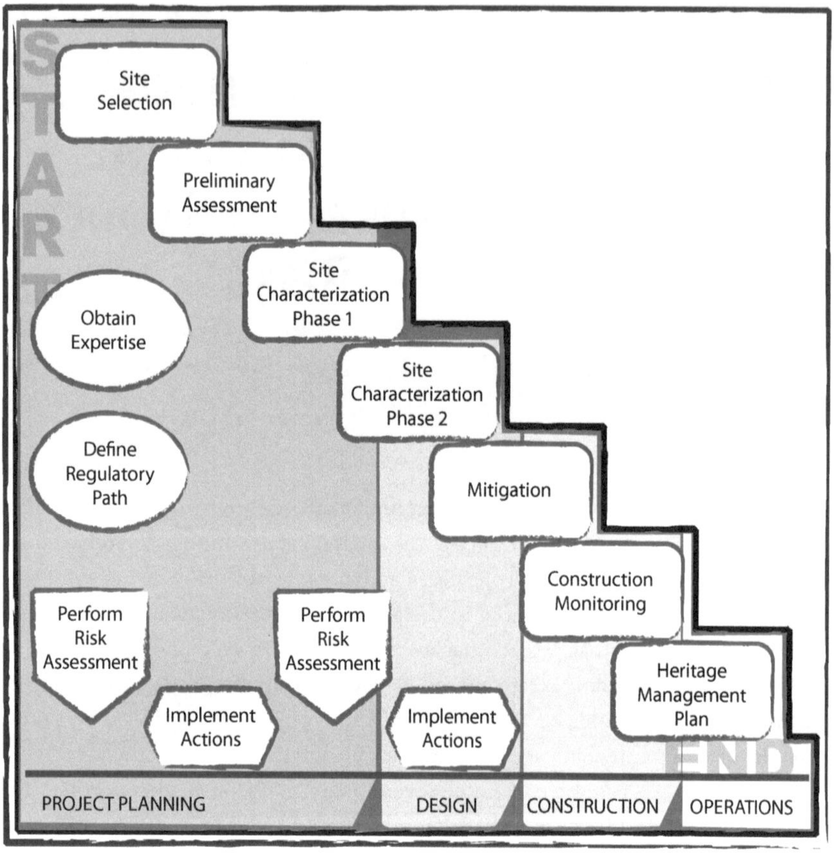

FIGURE 4.1. *Archaeological assessment steps and associated tasks from project planning through operations. As project activities begin planning efforts, it is important for the project to obtain expertise, define the regulatory path, and begin performing archaeological risk assessment.*

IDENTIFYING IMPORTANT ARCHAEOLOGICAL RESOURCES IN THE PROJECT AREA

Any archaeological resources that may exist in the project area should be identified as early as possible, and certainly before construction begins. The first effort conducted as part of the initial archaeological assessment will have provided some insight into the question of site potential. If people are in agreement that archaeological fieldwork is required, then some kind of subsurface testing strategy will have to be deployed. Identifying resources is usually not a simple matter. The right methods and sampling strategies need to be used if the results of the assessment are to be meaningful.

For example, it is critical to conduct sufficient background research to determine what may have been located in the project area in the past. It is common practice for a due diligence review to be conducted before a property is purchased, which provides some information about past use of the area. But these reviews are primarily concerned with more recent industrial uses. To supplement these reviews, one needs to look into a variety of other records to see if any archaeological sites have been located in the area, if there was historic settlement in the area, and if anyone ever lived on the property. Aerial photographs can also help.

As the project starts to take on momentum, it is time to begin site-specific, detailed investigations. Site investigations can be costly, especially when they involve subsurface testing, so it is important to develop questions that can be answered through the field investigations. The challenge for archaeologists is not unlike those facing environmental engineers who attempt to characterize the subsurface environment or subsurface contamination (Boulding and Ginn 2004; Nielson 2006). These questions may have been identified in the initial risk assessment, but if not, they should be developed prior to any archaeological testing. Targeted field investigations designed to verify the conceptual models provide good quality information useful in managing risks.

The challenge for the archaeologist becomes developing a good sampling strategy for the project area. The sampling strategy needs to be sufficient to confirm the presence or absence of archaeological deposits in different sections within the project area. Archaeological sampling approaches generally involve a combination of techniques:

- *Surface surveys:* Archaeologists walk over the project area, usually in transects 10 to 20 meters apart, looking for artifacts on the surface, changes in the surface associated with past activity, and vegetative patterns that may be associated with past use. Survey techniques may vary with landforms present within the project area.
- *Subsurface testing:* To collect subsurface evidence, archaeologists use different methods. Shovel tests can involve round holes, surface scrapings, or small square holes with material screened through a 1/8- or 1/4-inch screen. Other testing strategies involve excavation of 1-meter squares, 50-cm squares, and backhoe trenches.

There are a variety of archaeological techniques used around the world to accomplish the characterization efforts. Descriptions of archaeological techniques can be found in King (2005), Neuman and Sanford (2001), and White and King (2007). In addition to archaeology's more traditional, labor-intensive techniques, there are now high-tech approaches as well: remote sensing technologies such as satellite imagery, infrared aerial photography, ground-penetrating radar, electro-resistivity, and magnetometry. Use of these different technical approaches involve trade-offs between cost and quality of information obtained. Selecting the best combination of techniques is best left to professionals familiar with the types of landforms and resources expected in the project area.

The project manager should discuss the chosen strategy with the regulator and others who have expressed an interest in the project, and seek agreement on the sampling techniques to be used. There are too many cases where a project manager and archaeological contractor sampled a project area only to find out later that the regulator had a different idea about the work that should be done. The result: they had to resample the area.

If archaeological or historic resources are found within the project area, the next step involves determining whether or not the resources are important, and if so, why and to whom. This calls for a professional assessment that identifies the resource, analyzes its condition, and determines why it is or is not important. It is imperative that any professional assessment consider the views of others who may care about the resource. If the resources are present but not considered important, approval will be granted for the project to proceed. If the resources are deemed important, efforts will need to be made to avoid or mitigate the effects of the project on the resources.

MITIGATING EFFECTS

If important resources will be impacted by project activities, something needs to be done to minimize the impacts. Projects often will be redesigned to avoid an important resource; for example, the routing for a proposed pipeline can be changed, or a building can be moved. If redesign is not an option, then the archaeological deposit may have to be removed using archaeological data recovery techniques. Other resources, such as cemeteries, may be deemed so important that removal is not an option.

Data recovery involves large-scale excavations, which unearth large numbers of artifacts, each of which must be carefully removed, cleaned, sorted, described, analyzed, and reported upon. It is a slow process, labor intensive and expensive. The analysis and reporting can take four times as much time as the excavation. The details concerning the archaeological excavations—the area to be excavated, the data recovery strategy, analytical methods, consultation requirements, report requirements, etc.—will be detailed beforehand in an agreement document of some form (e.g., a memorandum of agreement, treatment plan, or excavation plan). These are legally binding documents which, while they may take a long time to develop and get signed, clarify the work to be done, which in turn allows their details to be incorporated into the project planning, specifically the schedule and budget.

Archaeological excavations are not always necessary. Other forms of mitigation can involve making public presentations about the cultural resources, through lectures, pamphlets, or books. For heritage resources such as buildings and engineered structures, the preparation of detailed archival-quality drawings and documentation is routinely required. Mitigation for impacts on landscapes and traditional cultural properties can involve agreements to minimize visual effects or to limit access and use.

It is not unusual for there to be disagreement over the protective actions that should be taken. The project lead may think one thing, the regulator another, and a stakeholder still something else. It is common for project personnel to spend more time and money fighting the recommendations of others than it would take to simply address the request.

Archaeological Monitoring during Construction

Whether or not to monitor for archaeological resources during construction is one of the most important decisions that must be made by those involved in the project. Archaeological monitoring entails having someone with archaeological expertise watch over construction activities, examining subsurface deposits as they are encountered to make sure archaeological resources are not present. As shown in Figure 4.2, there are multiple times during the course of a project when archaeological work must stop and the parties must decide whether to do more archaeological work, monitor construction activities, or do nothing more. Typically, the archaeological evidence suggests that no additional archaeological resources exist in the area, but because of inherent uncertainties, a recommendation to monitor when construction begins is often made. The advantage is that the agreement to monitor allows a project to stop site characterization activities; the disadvantage is that if archaeological resources are discovered during construction, as occurred at Port Angeles, impacts to the project from work stoppage can be substantial.

The decision must balance the cost of doing more archaeological characterization against the risk of a discovery during construction. If there is nothing important to be found, then any additional archaeological work is a waste of time and money. If something is there but has yet to be discovered, then the cost of discovery during construction far outweighs the cost of more characterization work. It is a judgment decision, but that determination should be based on the risk as determined by the risk management process outlined in Chapter 5.

In cases where there is archaeological mitigation through recovery of the resource, there comes a time when mitigation stops, per the agreement, and again, the parties must decide whether to monitor during construction or not. If the archaeologists are confident that the entire archaeological deposit has been removed from the ground, there is probably little reason to monitor.

It is our opinion that archaeological monitoring during construction should only be agreed to when there is a low probability for resources to be found.

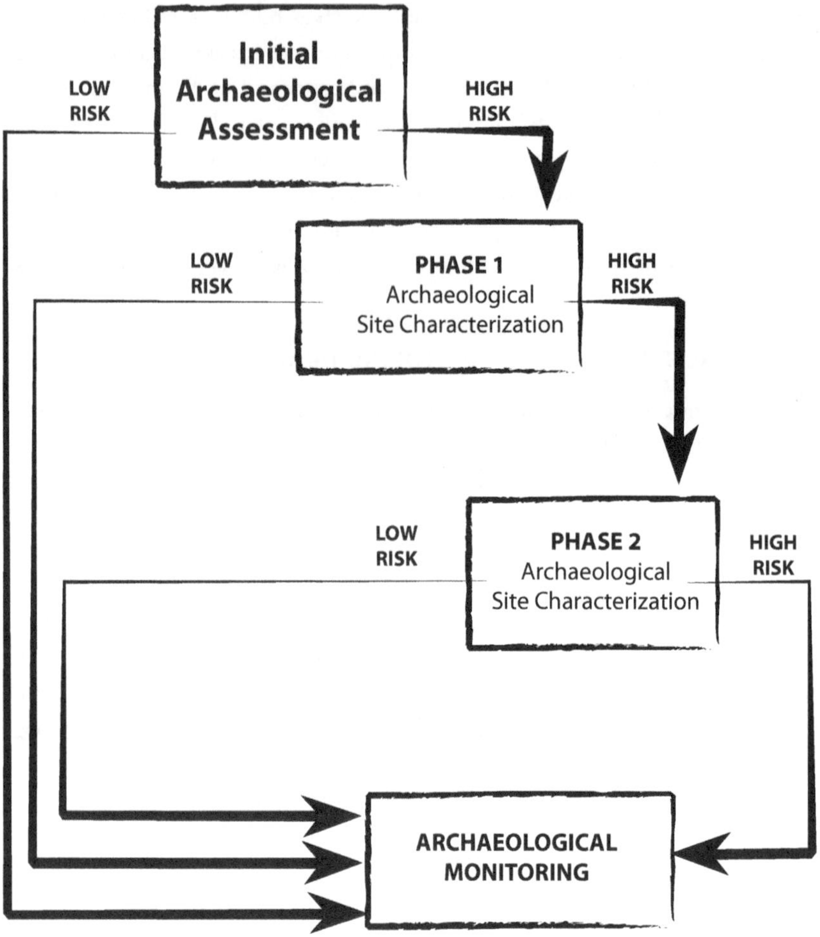

FIGURE 4.2. *The archaeological monitoring decision-making process. At any point in the process, project participants may decide to stop conducting more archaeological investigation and either cease all efforts or commit to archaeological monitoring during construction. It is critical that any decision to rely on construction monitoring be made only when the risk of an archaeological discovery is determined to be low.*

ADDRESSING INADVERTENT DISCOVERIES

Inadvertent discoveries present by far the worst scenario, especially if they involve human remains. If there is such a discovery during construction, regardless of whether a monitor is present, work will have to stop until the discovery has been evaluated. Work stoppages run up costs for construction personnel who are onsite and have nothing else to do. Most of the project design will have been done by the start of construction, so if changes are needed, additional costs will be incurred. Since time is of the essence, there will not be adequate time to consult and evaluate the situation and develop an appropriate and sensitive resolution. Resources have been damaged and cannot be replaced. People will likely be upset and angry; there will be negative publicity and, very likely, lawsuits. While all inadvertent discoveries may not be preventable, experience shows that most are the result of inadequate risk management, and thus were preventable.

It is important to be explicit about the process to be followed if a discovery is made. There needs to be a clear chain of command; the actions that will be taken need to be clearly identified. The individuals who must be informed of the discovery need to be identified, as do the individuals responsible for doing the notifying. The process for stopping work, performing assessments, and resuming work should be tied to timeframes.

DISSEMINATING FINDINGS

When important information has been discovered as a result of archaeological investigations, the information needs to be documented and distributed to the appropriate stakeholders. Some information is appropriate for the public and can involve various types of presentation—tours, talks, demonstrations. Some information is important for scientific research and needs to be documented and synthesized in peer-reviewed scientific journals. Some information may be relevant and possibly sensitive for descendants, and their interests should be considered first.

SUMMARY

In this chapter, we have presented the fundamental aspects of archaeological risk management: determining if important archaeological resources are located in the project area and what to do if they are. We discussed the need to adequately characterize the project locations through the use of archaeological sampling techniques and conceptual models. Also covered is the need to adequately mitigate the effects of the project on the resources before recommending that archaeological work stop and archaeological monitoring of construction begin. We closed with discussions on inadvertent discoveries and the need to disseminate heritage resource findings.

5 The Risk Matrix: A Tool for Managing Archaeological Risks

In Chapters 3 and 4, we presented basic concepts of archaeology and heritage resource management. In this chapter we present a tool to assist project managers in determining the work that needs to be done and the timing of that work. The tool guides a project manager through the archaeological identification, evaluation, and impact assessment efforts. It also helps to identify and prioritize potential problem areas and to develop actions that can be taken to keep problems from getting out of hand.

The Archaeological Risk Assessment Process

The archaeological risk assessment process entails three steps (Figure 5.1):

Step 1: Perform a baseline risk assessment. This step involves developing archaeological scenarios specific to the project, which means envisioning various possibilities such as discovering archaeological resources in the project area, not getting a regulatory permit on time, and learning that certain groups are reacting negatively to the project.

Step 2: Identify the actions that need to be taken to address the problems posed by those scenarios, and document the basis for such actions. This, combined with the scenario descriptions developed in Step 1, constitute a risk management report for the project.

Step 3: Revisit the risk scenarios regularly to determine if any new information has surfaced that affects the likelihood or consequence ratings. When new information warrants a change in ranking, it is time to reassess the scenario.

Each step is described in greater detail below.

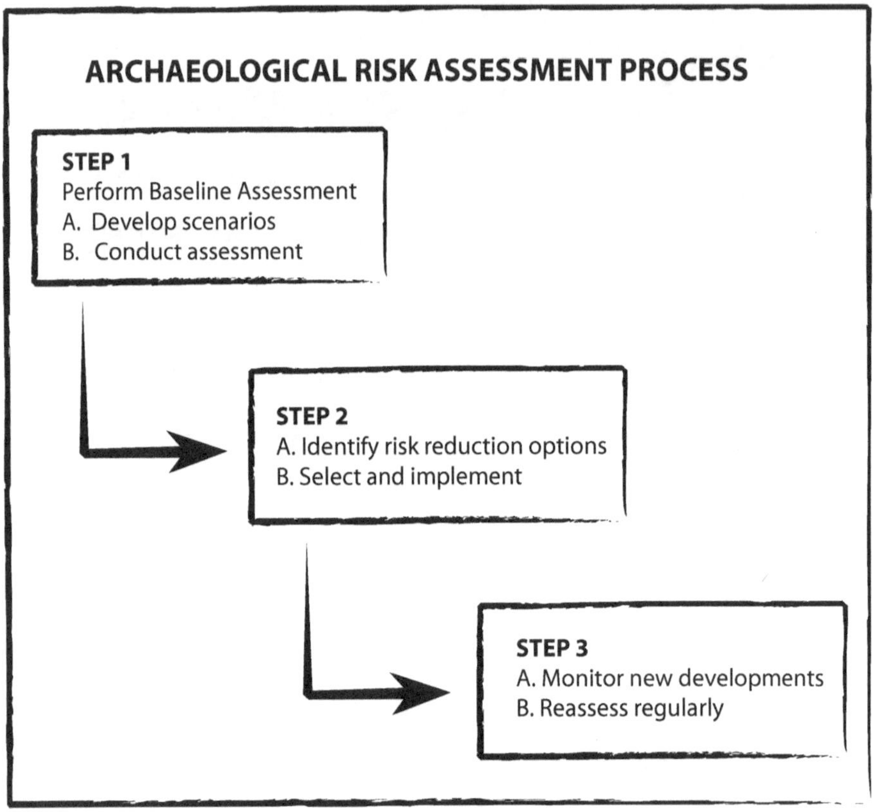

FIGURE 5.1. *The archaeological risk assessment process assists project managers in addressing archaeological issues, concerns, and expectations.*

Step 1: Perform a Baseline Risk Assessment

A baseline risk assessment should be performed early in the planning process. There are two major benefits: (1) the assessment will identify the critical issues that could arise in the project and allow project personnel to start addressing them, and (2) the initial assessment provides a baseline from which future assessments can be measured.

A risk management matrix is one way to help organize archaeological information associated with the project (Figure 5.2). The matrix presents the probability (or likelihood) that something (an undesired "risk event") will occur and the severity of the potential consequences (e.g., increasing costs, project delays, negative publicity, loss of community heritage) if it should occur. To use the matrix, planners ask: What can go wrong? How likely is it to occur? What are the impacts to the project if it does?

Develop Scenarios

To begin the process, scenarios applicable to the specific project need to be developed. The scenarios may be obvious, or they may require some brainstorming among project personnel. Below we list four common scenarios that fit with the Port Angeles and Blaine projects; we use these to explain the archaeological risk assessment process.

- The permit to proceed with the project may not be issued before the scheduled start of construction.
- There are groups interested in heritage resources that may be affected by the project.
- There are buried archaeological deposits within the project area.
- Archaeological deposits will be discovered during construction.

Each risk event is then assigned an initial likelihood and consequence ranking.

Determine the Probability of a Risk Event Occurring

The process of determining risk involves asking a series of explicit questions at the beginning stages of the project. Each scenario has its own set of individualized questions. Consider the regulatory issue: "The permit to proceed with the project may not be issued prior to the scheduled start of construction." To determine the likelihood

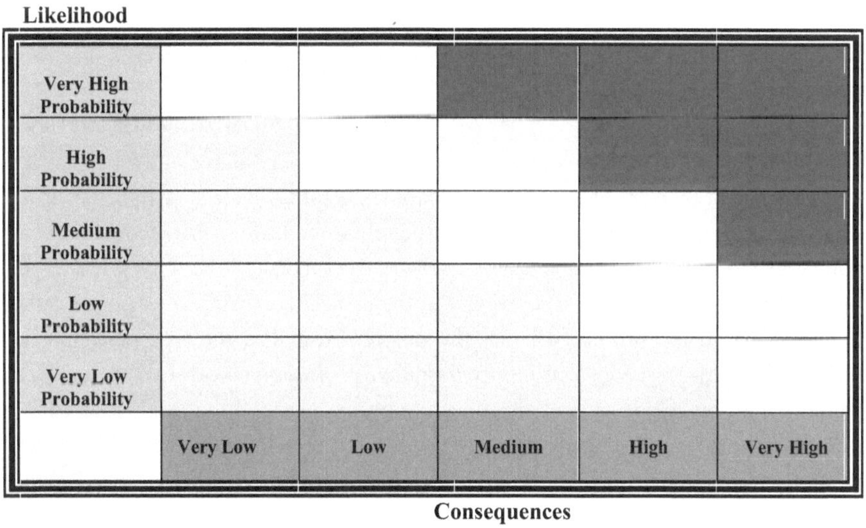

FIGURE 5.2. *A risk assessment matrix can be used to predict the likelihood and consequences of an event and the associated risk. Where probabilities and consequences are high (shaded boxes), actions need to be developed and taken.*

that the permit will not be received in time, the first question you might ask is: How long will it take to do whatever has to be done to comply with the regulations, and get the permit? As you begin to analyze the situation, you learn that the compliance work will take about three months, assuming no resources will be impacted by the project, but could take as long as a year if there are important resources in the area. If the project is scheduled to start construction in nine months, and you know there are important resources in and around the project area, you can say there is a Very High Probability that the permit will not be issued by the construction start date. Alternatively, if construction is a year away, and from what you have heard, the project is not in a sensitive area, then you could say there is a Low Probability that the regulatory process will stall the project.

Or consider the consultation scenario: "There are groups interested in heritage resources that may be affected by the project." To determine how likely it is that there are such groups, we might ask: Are there people with strong ties to the project area, who care deeply about it and who will want to know about the project? Suppose the project manager knows there are groups to consult with, but she has yet to contact anyone. We would assign this scenario a Medium Probability because we have a reasonable expectation that someone may have concerns. Or, suppose the project manager has made contact with various parties, based on recommendations received from the government archaeologist. The project manager has heard back from a few groups that they are interested in the area, but they have no concerns about the project. We would assign this scenario a Medium Probability, because we can be reasonably certain that the group will want to become more involved if something is discovered once construction begins.

For a final example, we consider one of the most important issues, the likelihood that there is an actual archaeological site or other heritage resource in the project area. Consider the scenario: "There are buried archaeological deposits within the project area." At the beginning of a project, there is likely to be little archaeological information collected about the project area, but the project manager should be able to get a general idea about the cultural sensitivity of the project area with a few phone calls. A critical question to pose could be: Is the project located in the type of area where archaeological deposits are commonly found? A professional with experience in the region should be able to answer this question, with caveats. If the site is located away from water in an area with little documented history, then the answer will be "not likely," and a reasonable probability score will be Low Probability. If the answer, "it's possible," then we'd give it a Medium Probability score. If the answer is "the project is located in a very sensitive area," then a High or Very High score would be appropriate.

Scores need to be revisited regularly, perhaps monthly, and especially at key points in the project. Suppose that during the initial risk assessment, your team determined there was a Medium Probability that resources would be discovered in the project area. Now you are planning to start the regulatory permitting process, so you want to get a more refined estimate than the one done during the initial assessment. To get an estimate that you can have confidence in, you need to obtain the services of a professional archaeologist.

For the initial assessment, the archaeologist would research what was known about the history of the area, perhaps checking historic documents, such as old maps, or asking elders in the local community what they know about the area. He or she might use analytical techniques or GIS software to predict where archaeological sites may be located (known as a *predictive model of archaeological site location*); often such analyses are based on statistical relationships between known sites and environmental factors such as distance to water, type of landform, prehistoric or historic disturbance of the landform, and distance to other sites. Depending how much money is available for the initial assessment, the professional will go to the project area, do a walkover, a systematic surface survey, or even conduct some subsurface testing.

Using the matrix presented in Figure 5.2, the probability of occurrence is assigned to one of the risk matrix columns:

- Very High: There is virtual certainty that the scenario will occur.
- High: It is likely to occur (> 75%).
- Medium: It is moderately likely to occur (25%–75% chance).
- Low: It is unlikely to occur (< 25% chance).
- Very Low: It almost certainly will not occur.

These probabilities are often determined from management experience or professional judgment and can vary depending on the analyst's level of expertise, experience, and perspective. It is important to consider the objectivity of the individuals doing the ranking and the quality of their analysis.

Some people find it very difficult to assign rankings. However, it is more important to get the scenario on the matrix, document the basis, and start tracking, than it is to struggle over correct placement on the matrix. The project manager (or leader of the risk assessment team) will need to consider the arguments made by different parties and make a final determination.

Estimate the Consequences

To estimate the consequence of a particular risk event, the analyst must ask: What would be the impact on the project should the event occur? Archaeological consequences are typically viewed from a project perspective: additional costs, schedule delays, heritage loss, negative publicity, and antagonism from various stakeholders. The following consequence severity levels might be defined in a risk analysis:

- Very High: The risk event would have significant impact on the project, including long schedule delays and large budget escalations.
- High: The risk event would add significant costs to the project, and some changes would need to be made in the schedule or budget.
- Medium: The project would be impacted, but the consequences would be manageable if handled correctly.
- Low: There would be some minimal consequence.
- Very Low: There would be no consequence to the risk event.

As with determining the likelihood of occurrence, determining the consequences of an event involves a series of questions. Returning to the regulatory scenario used earlier, the question might be: What happens if the regulatory permit is not obtained by the scheduled construction start date? If the answer is that construction cannot begin until the permit is obtained, and that creates a problem (loan problems, weather problems, fines, worker or equipment availability), then the consequences are probably High or Very High. If the permit delay does not matter because the project is already significantly behind schedule, then that's a Low consequence.

How a consequence is perceived will vary with the perspective of the person doing the risk assessment. For example, if the scenario is concerned with finding archaeological resources within the project area, and the question concerns the consequences of a discovery, project personnel will tend to look at how the project will be affected—there will be delays, additional costs, and so on. The archaeological regulator, by contrast, may think of an inadvertent discovery in terms of loss of scientific information. A representative of the descendant community may frame everything in terms of the impact on the resource or on the descendants themselves. A local government official is most likely to be concerned with the economic or educational impact on the community. Overall, the project manager should consider all—direct and indirect—impacts to the project. One should also occasionally look at the best and worst outcomes for different scenarios to get a feel for the boundaries of the potential problem.

Establish the Risk

After estimating the likelihood of occurrences and the severity of consequences, we can begin to talk about risk. Risk is determined by combining the likelihood that the event will happen and the resulting consequence. A matrix is one tool that can be used to illustrate the risk. We like a 25-cell matrix, with five rankings along each axis (Figure 5.2). For some projects, where likelihood and consequence cannot be estimated with such certainty, a matrix with three columns and rows rather than five may be more appropriate. Risk events should be entered into the matrix, and the assumptions that were used to estimate likelihood and consequence should be documented.

Step 2: Develop and Implement a Risk Management Plan

Scenarios placed in the High Risk and Medium Risk cells call for action plans to be developed. Action plans identify actions that could be taken to reduce the risk to an acceptable level. If, for example, there is a high likelihood that regulatory compliance work will fall behind schedule, then the action might be to hire an archaeological contractor to help the project get the approval process started. Or if there is a high likelihood that a particular group is going to have significant concerns, the action plan may call for meeting with this group to understand its issues, concerns, and expectations.

The risk management report becomes the primary mechanism for documenting the baseline archaeological risks identified in Step 1. The report is essentially a compilation of scenario descriptions that have been developed (Figure 5.3). Each scenario should have an explanation for the likelihood and consequence rankings, and options for reducing risks. Documentation can be brief if explicit. The following information should be captured:

- *Risk Event Description*: Describe the issue that confronts the project.
- *Probability Assessment*: Describe the basis for selecting the current ranking.
- *Consequences*: Elaborate on the nature of the current consequence assignment and why you think the impact is very low, low, medium, or high, or very high.
- *Next Steps:* Describe the next level of effort that would increase understanding of the issue; describe the effort that the project will undertake; and explain the project strategy. To determine the next steps, the project manager needs to decide how to manage the various risks articulated in the matrix. Two options might be:
  - accept the current risk and take no new action for the moment;
  - decide that the risk is unacceptably high and develop actions that can be taken either to reduce the likelihood of the risk event, or to lessen its impact, or both.

  Decisions to pursue specific actions are then documented.

Step 3: Reassess Regularly

The risk assessment needs to be updated regularly, as mitigation activities proceed. Whether this is done monthly or quarterly depends on the size, complexity, and length of the project, or on whether, and how quickly, a particular situation starts to deteriorate. At a minimum, assessments should be conducted as phases of the project are completed.

Hypothetical Scenario Description for the Port Angeles Risk Assessment Report

- **Scenario Description:** Archaeological Materials Exist in Graving Dock Area
- **Scenario Background:** The project area is located in a culturally sensitive area (Port Angeles waterfront). The village of *Tse-whit-zen* was located in the area. The project area is covered with industrial remains from the last 100 years. Concrete covers much of the surface, limiting ability to see the surface. It is not clear where the original shoreline and beachfront would have been, whether it has been altered, or if it is buried beneath the industrial remains.
- **Likelihood: High**. According to our disturbance model, there is an industrial stratum overlying the historic surface. The shoreline is believed to be more or less where it is today, with the historic beachfront under the industrial deposits. Archaeological deposits associated with ethnohistoric or prehistoric settlements would be along the shoreline. Until we have a good subsurface sample and know more about the extent of historic and modern disturbance as it relates to the original shoreline, we have to surmise that resources exist.
- **Consequences: High**. If there are extensive archaeological deposits located beneath the surface, it could take some time to get agreements in place and conduct the archaeological work. The Lower Elwha Klallam Tribe will likely want to be involved and would be especially protective of any tribal-related resources that exist. It could be hard to meet the schedule. If discoveries are made, the schedule will likely be impacted.
- **Next Steps:** We will contract for an archaeological assessment and regulatory analysis.

Figure 5.3. *Hypothetical scenario description that might be found in an archaeological risk assessment report*

Once completed, new risk assignments can be compared with previous ones to determine if risk has improved or deteriorated. Identifying trends on the matrix is a major benefit of the reassessment effort; changes in rankings can be identified with plus signs (+) where they increase and minus signs (–) where they decrease. As the project proceeds through the project phases, we should see reductions in probability and consequences each time a reassessment is done. This would be confirmation that risk-reduction decisions are working and that the archaeological situation is under control. In contrast, any change toward increasing probability or consequences would be an indication that the archaeological situation may not be under control. Increasing risks should be addressed immediately and with vigor, because change toward higher risks may mean that some of the fundamental assumptions and models developed for the project may be wrong.

APPLYING THE MATRIX

To demonstrate how the matrix can be applied, we use the Port Angeles Graving Dock and the Blaine Wastewater Treatment Facility projects discussed in Chapter 2 as examples, with the benefit of 20/20 hindsight.

Port Angeles Graving Dock Project

Examples of scenarios applicable to Port Angeles are presented below. The risk assessment matrix is shown in Figure 5.4.

- **Scenario 1**. *The archaeological compliance permit will not be issued before the scheduled start of construction.* Recent experience elsewhere in Puget Sound had identified the Endangered Species Act as the regulatory challenge to be met. Archaeological compliance with U.S. heritage protection laws pursued a separate path. How long would it take to get clearance from the Washington State Historic Preservation Officer for the graving dock project to proceed? At the start of the project, a time frame of four months would be a conservative estimate; so we consider it unlikely that the project would fail to get clearance within the allotted schedule (**Likelihood: Low**). And since construction was not scheduled to begin until well after that, we would rank the consequence as low if the archaeological permit were not obtained at the end of four months (and assuming delays would be on the order of weeks or months) (**Consequences: Low**).
- **Scenario 2**. *There are groups interested in heritage resources that may be affected by the project.* In seeking to answer the question, "Are there any cultural groups with special ties to the project areas," any inquiry within the local or professional community would have quickly identified the local Lower Elwha Klallam Tribe as a descendant tribe with a strong interest in the area (**Likelihood: Very High**). In terms of consequences, the question could be posed, What are the positive and negative outcomes of active involvement by the tribe? The graving dock could certainly be a positive experience for the tribe, as they would like to benefit economically from the project, just as others would. Based on the experience at Blaine a few years earlier, however, it was well known that tribal authority extended to cemeteries and burials and that there was strong interest in archaeology as well. There could indeed be very high consequences from tribal involvement if the situation turned into another Blaine (**Consequences: Very High**).

- **Scenario 3**. *Archaeological deposits are located onsite.* At the start of the project, any initial attempt to address this scenario would have to conclude that there was a high to very high likelihood that archaeological deposits would be found in the project area. It was well known throughout the community that there had been a large and important Indian village in the general location (**Likelihood: High**). In terms of the consequences of resources being onsite, anything that could lead to a delay of more than a few months could have a major impact on the schedule to repair the Hood Canal Bridge. Archaeology and the discovery of burials anywhere in the Pacific Northwest could certainly cause delays while people worked to forge agreements and do the slow and meticulous work of burial recovery (**Consequences: High**).
- **Scenario 4**. *Unanticipated archaeological deposits will be discovered during construction.* It would be difficult to estimate the likelihood of this occurring, since no archaeological assessment had been made at the start of the project (**Likelihood: Moderate**). Consequences could be determined with some certainty, as it was well understood that the graving dock was on a critical path for the bridge repair. Any delay in the graving dock would be a delay in the bridge, with delays leading to cost

Occurrence

Very High Probability					*Scenario 3* Interested Parties
High Probability				*Scenario 2* Archaeology Exits	
Medium Probability					*Scenario 4* Construction Discovery
Low Probability					
Very Low Probability		*Scenario 1* Regulatory			
	Very Low	**Low**	**Medium**	**High**	**Very High**

Consequences

FIGURE 5.4. *Risk assessment scenarios and rankings for the Port Angeles Graving Dock project. Scenario 1: Regulatory concurrence will not be obtained on schedule. Scenario 2: Archaeological deposits exist in the project area. Scenario 3: Interested parties will be upset and require consultation. Scenario 4: Archaeological deposits will be discovered during construction and cause shut downs.*

escalation and increased risk to human health from unsafe bridge conditions (**Consequences: Very High**).

Blaine Wastewater Treatment Facility Expansion

The same four scenarios used in the Port Angeles example are used for the Blaine Wastewater Treatment Facility Expansion project (see Figure 5.5).

- **Scenario 1**. *The archaeological compliance permit will not be issued before the scheduled start of construction.* At the start of the project, the cultural sensitivity of the project area was known, even if not fully understood. It was known that the regulatory effort would not be trivial. Without really getting into the process, it would be hard to predict how long the process might take (**Likelihood: Moderate**). It was important to expand the wastewater treatment facility, but if the start date of construction slipped a few months because of delays in getting an archaeological permit, there would be minor impact(s) (**Consequences: Low**).
- **Scenario 2**. *There are groups interested in heritage resources that may be affected by the project.* The City of Blaine and the Lummi Nation shared common territory, so

Occurrence

Very High Probability				*Scenario 3* Archaeology Exists	
High Probability				*Scenario 2* Interested Parties	
Medium Probability		*Scenario 1* Regulatory			*Scenario 4* Construction Discovery
Low Probability					
Very Low Probability					
	Very Low	**Low**	**Medium**	**High**	**Very High**

Consequences

FIGURE 5.5. *Risk assessment scenarios and rankings for the Blaine Wastewater Treatment Facility Expansion project. Scenario 1: Regulatory approvals will not be obtained on schedule. Scenario 2: Archaeological deposits exist in the project area. Scenario 3: Interested parties will be upset and cause problems. Scenario 4: Archaeological deposits will be discovered during construction and cause shut downs.*

their likelihood of wanting to be involved was well known (**Likelihood: High**), and the potential consequences were understood to be potentially significant (**Consequences: High**).

- **Scenario 3**. *Archaeological deposits are located onsite*. That archaeological remains and a cemetery were located in the immediate area was well known, as they had been previously located and excavated during the original construction (**Likelihood: Very High**). If there were deposits in the project area, significant costs could accrue in a project that was already underfunded (**Consequences: High**).
- **Scenario 4**. *Unanticipated archaeological deposits will be discovered during construction*. It would be difficult to estimate the likelihood of this occurring. Archaeological deposits were known to exist, but there would also be archaeological work done prior to the start of construction, which should prevent anything being discovered once construction began (**Likelihood: Moderate**). Consequence could be determined with some certainty, as it was well known that the project was on a tight budget. Any discoveries during construction could lead to work stoppage costs and additional archaeological costs (**Consequences: Very High**).

With baseline information documented, a project can begin to develop and implement a risk management plan.

ARCHAEOLOGICAL RISK ASSESSMENT REPORT

The information collected and analyzed is documented in an archaeological risk assessment report. The archaeological risk assessment report provides the basis for decisions made concerning archaeological resources and identifies places where additional information may be needed. The report, which builds on the initial cultural assessment, typically includes such sections as:

- Project History, Requirements, and Consultations
- Environmental History of the Project Area
- Cultural Context of the Locality and Project Area
- Evaluation of Known and Suspected Heritage Resources
- Description of Heritage Site Conceptual Site Model(s)
- Predictions for Use in Assessing Archaeological Risk
- Next Steps

Each section is briefly discussed below.

Project History, Requirements, and Consultations

This section needs to include a basic description of the project, especially those activities that will be ground disturbing or have visual impacts. Requirements that should be addressed include alternative ways to comply with regulations, and identification of consultations that must or should be done.

Environmental History of the Project Area

A good description of the environmental history is needed to help identify those areas of the site that have potential for heritage resources. This description should also identify areas where there is no potential, either because the landform has been disturbed from previous construction, or because the sediments are too old to contain resources. Other environmental aspects of the project area, such as vegetation patterns and surface water distribution, can provide the basis for the environmental description.

Cultural Context of the Locality and Project Area

Any previous archaeological, historical, and cultural research that has been done in the general area needs to be synthesized so that settlement patterns for the different time periods can be identified. This information will also be useful for identifying people it may be important to consult and places where you may find additional information.

Learning about the cultural history of the project area is critical to good archaeological risk management. Researching information repositories for existing data is important work, but it is also important to talk to people familiar with local history and who can identify other people and places to go to for additional knowledge. Background research takes time, but it is worth it. A good grasp provides a solid foundation for managing the archaeological risks.

Evaluation of Known and Suspected Heritage Resources

If there are known or suspected heritage resources in the project area, these should be described. Included in the description should be an explanation of why the resources are or are not important, and to whom. Many times, the project area will be divided into zones based on cultural sensitivity.

Description of Heritage Site Conceptual Site Model(s)

This is perhaps the most important part of the assessment report because it identifies the assumptions that have been made about the project area. The description should clearly identify the resources that are either known or predicted for areas of the project, and the basis for those descriptions. Where competing models develop, additional data should be collected that will either confirm or reject one of the models.

Predictions for Use in Assessing Archaeological Risk

This section provides the rankings that are used for risk event likelihood and consequences in the risk management matrix. The likelihood rankings should be based on the project area conceptual model and the degree of certainty that exists. The estimate of consequences for the various scenarios should be based on what has been learned through the various consultations. An indication of certainty should be provided with each estimate.

Next Steps

This section identifies the next steps that the project team has decided to take concerning the identified risks. Where risks are high or increasing, activities should be identified that will stabilize or reduce risks.

SUMMARY

In this chapter we have presented a risk management tool, the matrix, to assist project managers in balancing the probabilities and consequences of specific risk events. When the risks are too high, the project team develops actions to reduce those risks. This approach is similar to many standard risk reduction approaches used around the world.

The tool can also be useful for others involved in the project—perhaps a representative of the descendant community or a regulator. These representatives can also estimate the likelihood and consequence of a scenario from their own perspective, which may be different from the estimates of the project. The matrix, then, becomes an effective way for different parties to develop and discuss scenarios and explore ways to manage the associated risks.

6 CASE STUDIES

In this chapter we present three case studies, two of which were prepared by colleagues with whom we have had many conversations about preventing archaeological disasters. The first case is one of the classic North American archaeological disasters: the African Burial Ground. The following description was prepared by Tom King, whose publications (2001, 2003, 2005, 2007, 2008, 2009) have shaped a generation of American cultural resource professionals. Tom knows this case well, and so we asked him to summarize the events and explain them for this book.

The second case is from the Missouri River in South Dakota, United States. A federal agency wished to construct a public boat launch, but failed to do a thorough historical and cultural review of the project area beforehand. The agency also failed to take seriously the protests of local American Indians who were concerned about sacred remains that were located there. Michael Burney, whom we have worked with for 20 years (Stapp and Burney 2002; Burney and Van Pelt 2003), was involved in this case and provided his account for the book.

The third case is one with which we are familiar. In this case, the disaster would have been the loss of an important archaeological collection representing a major American World War II site. Here the archaeologist not only failed to recognize the importance of the discovery, but also failed to listen to the concerns of those most interested in the resources that were discovered. The case was especially controversial because of the age of the archaeological deposits. One side argued that society already knew everything about the people and events that occurred at the site in the 1940s, and thus the site and its artifacts were "interesting" but not "important." The other side saw the site as a storehouse of new information that could be combined with documents and oral histories to further tell the story of an important chapter in world history

As you read through the cases you will recognize many of the issues discussed in previous chapters: the need to adequately characterize the project area for archaeological resources before construction begins; the need to consult with those who have an interest in the archaeological deposits; and the need to use qualified archaeologists.

Think about what the risk matrices might look like as each project proceeds; we include example assessments for each case.

DISASTER ON BROADWAY: THE AFRICAN BURIAL GROUND
by Thomas F. King

In the early 1990s, the U.S. General Services Administration (GSA) began work on a new federal office building on Broadway in New York City, a short distance north of Wall Street—in other words, a short distance outside where the perimeter wall of colonial New Amsterdam once stood. The project was on a fast track, under a design-build contract.

That contract was the first problem. The planning and construction schedules were essentially merged. It was assumed that there would be a seamless progression from design to build. This works only if you're sure that there will be no obstacles to progress.

GSA *was* sure, but it shouldn't have been. That was the second problem. The design-build philosophy doesn't envision that environmental impact assessment could actually unearth serious, project-halting impacts; it's just another bar on the Gantt chart; something to get out of the way so the project can move forward. That was the way GSA viewed review of the project under the National Environmental Policy Act (NEPA) and Section 106 of the National Historic Preservation Act. They did an environmental assessment (EA), but never imagined that the EA would turn up significant impacts. They executed a memorandum of agreement (MOA) under Section 106 that focused almost exclusively on the new courthouse's visual impacts on an adjacent historic courthouse.

The EA research turned up a map from the early eighteenth century. It showed a pond on what was now the project site, called the "Collect Pond." On the edge of the Collect Pond was something labeled "Negro burial ground." Other documents indicated that this location, outside the colony's wall, was where enslaved Africans and African Americans buried their dead.

GSA's reaction was the third problem. The agency treated the burial ground as something that archaeologists could make go away without disrupting the construction schedule.

This cavalier attitude in part reflected another mistake—failure to think closely about the project site's geomorphology. Water bodies like the Collect Pond form in low places—that's hardly rocket science. Hence the burial ground as of the turn of the eighteenth century was also in a relatively low place, surrounded by rolling terrain. Manhattan today may rock, but it doesn't roll; it's flat. The area has been leveled, so formerly low areas are relatively deeply buried today. GSA didn't take these elementary facts into account; it assumed that the basements of buildings constructed over much of the site in the nineteenth century would have wiped out most of whatever was buried there. GSA figured that the archaeologists probably wouldn't even have much to uncover.

And because it saw the burial ground as simply an archaeological problem—and because its regional office was coordinating with the office of then-Mayor David Dinkins, who was of African ethnic origin—GSA saw no reason to consult the city's African-American community. This was the ultimate mistake that set the project up for disaster.

The construction contractor was ready to go to work when the archaeologists began to find coffins and skeletons. First one, then several more, then a bunch more. And more, and more—because, of course, the nineteenth-century basements had not been deep enough to wipe them out. Eventually over 400 bodies were found.

News of the discovery got out; African Americans and their supporters began to protest, hold press conferences, picket the site.

It was about at this point that GSA hired me to help its national office with archaeology and historic preservation. Still trying to treat the burial ground as an archaeological matter, the regional administrator in New York asked me if they couldn't sample it somehow, to reduce the cost of digging up all those bodies.

"Well," I said, "maybe you can imagine how African Americans are going to feel about the bones of their ancestors being shoveled into dump trucks. But if you consulted with them and they agreed, maybe you could excavate a sample with archaeological controls and dig the rest of them up quickly, like coroners do." GSA jumped on this notion, but since the idea hadn't emerged from consultation with the affected people, the proposal to use a "coroner's method" only inflamed the situation further.

The project became the subject of a congressional investigation, worldwide protests, and extremely negative publicity and eventually was shut down. After much negotiation, it was allowed to go forward, reconfigured to reduce damage and with an extensive, expensive program of archaeological and anthropological research—which has had fascinating results. Part of the project site sat for years as a vacant, fenced lawn, but now there's a dignified memorial managed by the National Park Service. The site has been designated a National Historic Landmark and is the centerpiece of a substantial public education and research program. So good things have come out of the disaster on Broadway, but the cost to GSA was great—in angst, in damage to its reputation, and in taxpayers' money. The African Burial Ground website cites a cost of $50 million, but long before design work even began on the memorial, GSA financial officers were calling the vacant project site "our eighty million dollar lawn."

How could this disaster have been avoided? First, by taking environmental impact analysis seriously, and paying attention to the results. Second, by making the assessment transparent and consultative. As soon as the old map was discovered, GSA should have started consulting African-American interests about how to handle any graves that might turn up. Had they done this, the archaeology might not have been much less expensive, but it could have been better scheduled and done in a more orderly way—perhaps without seriously delaying the project. Third, GSA should have paid attention to the geomorphology of the site and its implications for the preservation of things underground, like bodies.

The case of the African Burial Ground didn't have to be a disaster, but it was, largely because GSA didn't take its environmental impact assessment responsibilities seriously and didn't consult properly with stakeholders.

EXAMPLE SCENARIOS—AFRICAN BURIAL GROUND

DURING PLANNING PHASE

Scenario 1: Important resources are in project area

- *Likelihood*: High (Negro burial ground on historic map)
- *Consequence*: Moderate (not known what impact would be)
- *Action*: Commission archaeological assessment to include geomorphology description, archaeological testing, and interviews with interested parties

Scenario 2: There will be significant delays because of permitting

- *Likelihood:* Low (environmental assessment proceeding on course)
- *Consequence*: Moderate (significant costs from project delays)
- *Action:* None.

UPON DISCOVERY

Scenario 1: Important resources are in project area

- *Likelihood:* Very High (burial ground has been located)
- *Consequence:* High (recovery work will be required, slowing project)
- *Action:* Begin discussions with state archaeology office and local groups; commission archaeological assessment and recommendations

DISASTER ON THE MISSOURI: THE YANKTON SIOUX AND THEIR BELOVED WHITE SWAN

by Michael S. Burney

The failure of anthropologists and archaeologists to adequately assist American Indians can and does have adverse effects on native people: spiritual sickness, physical calamities, social chaos, and financial stress. Indians perceive sacred places, artifacts, and human remains much differently than most non-Indians, including many anthropologists and archaeologists. American Indians, of course, have never approved of the destruction of their villages, traditional food-gathering areas, fishing grounds, or sacred areas—including places where their loved ones are buried. That fact has not changed in the modern era. Archaeological excavation destroys sites in order to "preserve" them, and incomplete background research by archaeologists only serves to provoke the Indians. Destruction of human burials and archaeological resources may result.

A case in point is White Swan, South Dakota, which comprises two places significant to the Yankton Sioux Tribe (*Ihanktonwan*): (1) White Swan (*Mága-ska*), within the White Swan State Recreation Area, and (2) North Point (or, as Yankton Sioux tribal elders prefer, East White Swan), within the North Point State Recreation Area. White Swan and North Point are on the east side of Lake Francis Case (the Missouri River—*Mini Sose Wakpa*, or Turbulent River) near Fort Randall Dam and Pickstown (*Wíyannunpapi*, or Double Woman) in Charles Mix County, South Dakota (see Burney and Scarlata 2003). According to the Yankton Sioux, they own both White Swan and East White Swan, as guaranteed under the 1868 Fort Laramie Treaty. The State of South Dakota and the U.S. Army Corps of Engineers, Omaha District, chose to disagree with the Yankton.

This desecration of Mother Earth cradling contemporary native peoples' ancient cultures distresses the Indians. Disenfranchised people are always at a disadvantage, and Indian tribes are no exception. In the White Swan case, the Yankton Sioux Tribe was unnecessarily forced into expending significant amounts of labor, funds, and other scarce resources to defend their culture, dignity, sovereignty, and, most significantly, their ancestors. The local and national press issued news stories sympathetic to the Yankton's plight, but Corps archaeologists ignored the coverage: not a single news article was included in the two brief archaeological inventory reports. As has often been the case, archaeologists' hasty surveys can adversely affect people's lives. In this instance, it was the Yankton Sioux who were wronged. But ultimately, both the tribe and the government were affected by these slapdash surveys.

Disregarding the Yankton Sioux's vehement protests, the Army Corps of Engineers transferred the North Point State Recreation Area to the State of South Dakota on or about February 19, 2002, although the Corps did agree to retain responsibility for archaeological matters (i.e., Section 106). On the surface, this sounded reassuring, but ironically, it was archaeology itself—or more precisely, lack of archaeology—that outraged the Yankton Sioux Tribe.

The State of South Dakota was anxious to expand the recreational capabilities of the North Point area—an already well-documented and previously disregarded historic area—by adding new roads, parking lots, comfort stations, dumping stations, and approximately another hundred camping spots. Corps archaeologists assessed the area as (1) devoid of cultural resources, (2) lacking the probability of finding further cultural resources, and (3) ineligible for the National Register of Historic Places.

Nothing could have been further from the truth. On May 14, 2002, workers moving dirt at East White Swan uncovered the remains of a young Indian woman and her two children, with associated funerary objects in their shallow graves (Figure 6.1). Thus began yet another emotionally wrenching indignity on the part of the Yankton Sioux, who simply wanted their "archaeology," and their dead, left alone. Instead, and without notifying the Yankton Sioux Tribe, the remains were quickly transported to the State Archaeological Research Center (SARC) in Rapid City, under authority of the Native American Graves Protection and Repatriation Act of 1990 (NAGPRA). Stunned tribal members reacted angrily, demanding their ancestral remains be returned.

White Swan, now underwater after having been submerged beneath Lake Francis Case—a result of the 1944 Flood Control Act—is a multicomponent prehistoric and historic site at least 160 acres in size. Ancient Woodland-aged occupations (A.D. 1–1000) were discovered there. The site also includes a traditional Yankton Sioux burial ground, the historic White Swan community, St. Philip's Episcopal Church and cemetery, the St. Francis Catholic Church and cemetery, private residences, a ferry landing, a stagecoach station, retail businesses, and more. In fact, White Swan was one of John Collier's "celebrated colonies." (Collier was Indian Commissioner from 1933 to 1945). Believing Indians would improve their lives by returning to their traditional communalism, Collier established four of these colonies on the Yankton Sioux Reservation.

Flooding of the Fort Randall Reservoir began in the summer of 1953, reaching full operating-pool level by the fall of 1954. The Pick-Sloan Plan—which turned the Missouri River into a series of reservoirs for power generation, irrigation, and recreation—devastated the archaeology and tragically disrupted the Sioux living on the Missouri River bottomlands. As many as 580 families were uprooted and relocated from their sheltered bottomlands to the higher windswept prairies—forcibly removed from their river, its life-giving resources, and their Missouri River way of life.

Low water levels in the reservoir reveal human remains, caskets, and burial items where the two churches and their cemeteries are located—common knowledge to collectors of Missouri River Indian relics and human remains. Present-day Yankton Sioux tribal members know their history along the river to a far greater extent than any anthropologist or archaeologist.

Ellsworth W. Chytka is one such tribal member, who provided extensive testimony in June 2002: "I have previously told . . . Corps of Engineer employees that Native American human remains are located at the North Point Public Recreation Area and that construction activities such as digging would cause disturbance of graves." Chytka was deeply saddened by the removal of large cottonwoods in preparation for the additional recreational development at White Swan. Cottonwoods are sacred; removal of these majestic trees only weakens the landscape and the Indians who rely on them.

Judge rules against Yankton Sioux Tribe

Sacred ground to become campground

By David Melmer
TODAY STAFF

PICKSTOWN, S.D. – The state of South Dakota has been given a green light to turn sacred burial grounds into a waste dump for motor homes and campers and a place to deposit fish entrails.

U.S. District Judge Lawrence Piersol issued his opinion that allowed the state to put back some dirt that was used for landfill to the location where a burial site was exposed, also that some dirt moved from another location could stay and be the foundation for the waste dump.

Nearly a year ago the state was excavating dirt from a location on North Point Recreation area when human remains and other artifacts were found. The Yankton Sioux Tribe went to court to stop the excavation, but now with the tourist season and the Lewis and Clark bicentennial approaching the work must be done, and the judge said the state can proceed.

"These are not just bones. These are human remains. This is sacred ground," said Ellsworth Chytka, spokesman for the Yankton Sioux Tribe.

"Where is there justice for us? When will the rights of humans win out over money? This is not just a Yankton matter. What needs to be done is all Native Americans in South Dakota defend their rights to the sacred sites and manage the water," Chytka said.

"It's time for the Great Sioux Nation to come together."

The state, in testimony at trial, argued the recreational site does not now have adequate waste

See **SACRED GROUNDS** *page* **B2**

FIGURE 6.1. *Archaeological disasters typically result in negative publicity. Events are often chronicled in local, regional, and national newspaper articles, such as this example from the White Swan case, which appeared in* Indian Country Today *(Melmer 2003).*

Natives long have seen archaeology as contentious at best. Only in the last 20 years have American Indians been made increasingly welcome in the archaeological profession in the United States. While archaeology has helped native peoples in numerous ways, regrettably, the White Swan debacle was not one of them. In the White Swan case, the archaeologist could have approached the situation differently. The Corps and the State of South Dakota had received enough advice, and had heard enough concern from the tribe, to justify the most cautious approach to any disturbance undertaken at White Swan. In fact, the Yankton Sioux's testimony alone was sufficient to justify undertaking more intensive fieldwork (e.g., archaeological testing and full-time monitoring).

Archaeology can lead any practitioner down a politically sensitive path. Regardless, it is vital to draw upon our anthropological and archaeological training. To start, a reasonable level of background research must be completed in order to fully document the anthropology and archaeology of an area. In the case of White Swan, the background research should have begun by consulting with the Yankton Sioux—and listening to them, not with wooden ears but with big hearts. Tribal members recall a time when White Swan and East White Swan were still viable communities. Many of their relatives once resided there and are buried in one of the two church cemeteries or in unmarked graves scattered along the Missouri River.

Yankton Sioux tribal members could have been employed to provide noninvasive archaeological testing; and tribal members could have provided monitoring services to early detect and minimize damage to buried cultural deposits, including human remains. The use of archaeological monitors is an effective way to minimize adverse impacts to unmarked buried cultural remains.

Workloads and busy schedules can keep archaeologists from undertaking in-depth background research and report preparation. This is a fact of life for many professionals in this field—especially Bureau of Land Management and U.S. Forest Service archaeologists working in Colorado, Montana, and Wyoming, who struggle to stay ahead of the frenzy of energy development. But busyness is no excuse. An archaeologist is *first an anthropologist,* with a professional responsibility to promote stewardship of present-day native peoples, their ancestors, and their archaeological record. Curiosity, research, and writing are fundamental tasks of academic, private, and government archaeologists, including those who participated in the emotionally charged North Point case.

Unfortunately, it was the failure of the Corps archaeologist to prepare a well-informed "case" that allowed the inadvertent exposure of native human remains. Corps archaeologists filed two "negative" cultural-resource inventory reports of several pages in length. A minimal records check undertaken with the South Dakota State Archaeological Research Center yielded one previous archaeological inventory and two Indian sites, neither of which could be located at the time of the surveys. Unconscionably, no mention was made of the prolific testimony provided by Yankton Sioux tribal members regarding the area's sensitivity. Nor were any of the local media findings represented.

The definitions for "literature search" and "records check" provided in *The Management of Archeological Resources*, edited by Charles R. McGimsey III and Hester A. Davis (1977), remain as relevant today as they were three decades ago.[1] Granted, not all archaeological projects warrant extensive background research, but many do; and the time and effort it takes frequently yields useful and valuable information that can save relationships, credibility, and money. This can especially be made possible when American Indians have an interest in the project.

A serious archaeologist would like to have as much information about an area as reasonably possible before beginning fieldwork. And the literature search and records check can be ongoing throughout the project. In the case of White Swan, after Burney & Associates was hired by the Yankton Sioux, the resulting report, "The Ihanktonwan's Sacred Place Called White Swan, in the White Swan State Recreation Area, and East White Swan in the North Point State Recreation Area, Lake Francis Case, Yankton Sioux Reservation, Charles Mix County, South Dakota," cited more than 300 sources (Burney and Scarlata 2003).

Archaeologists working on or near Indian lands would do well to listen to and work with the Indian tribes to the greatest extent possible. Indians, especially native populations residing in their traditional homeland—including present-day reservations—are often well informed about the surrounding landscape, historic use of the land, and traditional cultural properties, including previous camps, settlements, and burials. Yankton Sioux oral history and contemporary testimony supported the well-known fact that White Swan contained human burials as well as substantial other cultural resources, such as sites used for plant collecting, memorial gatherings (*wokiksuyes*), prayer gatherings, and for ceremonies like the Calumet ceremony (*hunka*), traditional pipe ceremonies, Sun Dances, and other essential tribal cultural activities. But the *nuh can* (wooden ears) would not listen to the *Ihanktonwan's* pleas and prayers. While Indians who provide this type of information may not have degrees in anthropology, or be practicing contract archaeologists, they are better informed than any outsider—including archaeologists. Understanding this fact inevitably leads to better outcomes.

It would have been a cautious, logical step to undertake archaeological testing at White Swan to detect the presence or absence of cultural material—including human remains—despite the negative findings of the two archaeological inventories. Dense vegetation covers much of White Swan, effectively obscuring cultural remains (including the two Indian sites hidden from the Corps archaeologist). Archaeological testing is limited in its effectiveness, depending on how much dirt is moved. Testing can be accomplished using hand tools, a small backhoe, or a combination of both, and should be rigorously employed by the archaeologist.

Archaeological monitors can be a cost-effective way to minimize adverse impacts from inadvertent discoveries exposed by mechanized equipment. It is baffling how the small expense of maintaining monitors on a project is routinely denied only to see the endeavor end up costing tens of millions of dollars more when archaeological materials, especially human remains, are unexpectedly exposed—an all-too-common scenario (Burney and Sampson 2004).

While it is impossible to calculate the total cost overruns due to cutting initial corners, suffice it to say they were far higher than necessary. Had the Corps archaeologist undertaken an adequate literature search and file check (one that included input from the Yankton Sioux), archaeological testing, and monitoring, she would have been hard pressed to conclude that White Swan was devoid of cultural resources, to ignore the *potential* for cultural resources, or or to say that this place was not significant and not eligible for listing in the National Register of Historic Places. Had the archaeologist shown an interest and curiosity about the written and oral testimony provided by Yankton Sioux tribal members and others, she likely would have been able to document the high probability for buried cultural resources, along with the importance of White Swan to the Yankton Sioux.

From the very beginning of a project, it is prudent to consult with the tribe and employ tribal members to consult on the overall design and implementation of the

EXAMPLE SCENARIOS—WHITE SWAN

DURING PLANNING PHASE

Scenario 1: Important resources are in project area

- *Likelihood:* High (historic occupation in area; tribal concern about burials)
- *Consequence:* Moderate (costs of resource recovery)
- *Action:* Commission archaeological assessment to include historical research, geomorphology description, archaeological testing, and interviews with tribes

Scenario 2: There will be significant delays because of permitting

- *Likelihood:* Low (regulatory approval expected)
- *Consequence*: Low (delay boat launch schedule)
- *Action*: None

UPON DISCOVERY

Scenario 1: Important resources are in project area

- *Likelihood:* Very high (burial ground has been located)
- *Consequence:* High (recovery work will be required, slowing project)
- *Action:* Begin discussions with state archaeology office and local tribes; commission archaeological assessment and recommendations

archaeology, to conduct archaeological test excavations, and to monitor the heavy equipment operation. It just makes good sense to work closely with and to involve the tribe or tribes that have a vested interest in the project. This is a small part of their story, their connection to the land, *mini sose wakpa,* and the historic cultural and spiritual place called *Mága-ska* or White Swan.

NEAR DISASTER ON THE COLUMBIA: SAVING THE MANHATTAN PROJECT LANDFILL

by Darby C. Stapp

One near disaster concerns a historic landfill associated with the construction of the world's first nuclear reactor, the reactor that produced plutonium for one of two atomic bombs used at the end of World War II. During exploratory excavations at a suspected contaminated site in 2005, cleanup crews encountered thousands of glass bottles, ceramics, and miscellaneous artifacts. The archaeologist who was there to monitor for possible intrusions into the adjacent prehistoric site did not recognize the historic nature of the material and failed to stop the backhoe trenching or to alert anyone after the fact.

When an archaeological crew from a local tribe happened by the area and saw the scatters of artifacts on the surface, they reported the discovery to the U.S. Department of Energy (DOE), which did not consider the discovery an archaeological site. The Washington State archaeologist, however, intervened and recommended that an archaeological site form be completed and a National Register evaluation conducted. The archaeologist who had originally ignored the discovery was asked to fill out the site form and write up the evaluation. He recommended the site not be eligible because it was heavily disturbed and the information contained therein not important, given the fact that extensive documentation and oral histories existed for the first reactors at Hanford, one of which was the B Reactor, a National Historic Landmark. The Washington State Historic Preservation Officer (SHPO) concurred.

One of the tribal archaeologists, however, recognized the archaeological site as extremely significant. Located about 400 meters from the Hanford Construction Camp, which was occupied between 1943 and 1945, she believed the artifacts were directly associated with workers at the camp who had built Hanford. Contrary to the views of the Hanford archaeologist and the Washington SHPO, she believed that the archaeological evidence, tightly dated to a three-year time span, was significant in its own right, but was made even more significant by the existence of documents and oral histories; further, her opinion was that the landfill deposits were intact, and not disturbed as the others had claimed.

A senior historical archaeologist came to her defense, and citing regulations (36CFR 800.4(c)(2)), requested that the Advisory Council on Historic Preservation ask DOE to seek an eligibility determination from the Keeper of the National Register. The Advisory Council made the request, and the project came to a halt while the documentation was prepared to submit to the Keeper. One year later, the Keeper agreed with the tribal archaeologist (and 30 others who wrote letters) and overruled the DOE and the Washington SHPO. News of the National Register decision made the front pages of the local newspaper. The decision vindicated the tribal claims and confirmed what most professional archaeologists saw as obvious: the site was extremely important. As of this writing, a memorandum of agreement detailing the mitigation actions that will be taken has been developed, and the archaeological work is scheduled to be conducted in late 2009.

How could such an obviously important archaeological site have come so close to destruction? Certainly we can point to the lack of expertise as a root cause. The archaeologist monitoring the work was not trained in method and theory of historical archaeology. He failed to recognize the significance of the site and failed to bring in available expertise to assist; in short, he gave his client bad advice. A second cause was the failure of the federal agency to adequately consult with other professional archaeologists and local tribes, as required by law; even when outside experts sent their concerns to the agency, their comments were not considered.

What were the consequences of this disaster? We estimate the agency spent approximately $100,000 fighting this case in staff time to attend meetings and prepare documentation. Professional reputations have been damaged. The management concerns of the Tribal cultural resource program were ignored, which led to additional concerns about the Hanford cultural resource program in general. The tribe was upset for being ignored. Furthermore, more time and money will still be needed before the cleanup can begin.

The schedule delays, additional costs, and conflict did not have to happen. Had the archaeological discovery been reported immediately, qualified professionals in historical archaeology would have gotten involved and determined the appropriate course of action. In this case, the initial disturbance of the site probably could not have been avoided. No one knew the landfill was there, and no amount of additional research would have led to its discovery prior to the backhoe trenching. As is often the case, however, the agency spent more money fighting than it would have cost to simply do the right thing. After five years, the archaeological work and cleanup is finally about to begin.

EXAMPLE SCENARIOS—MANHATTAN PROJECT LANDFILL

DURING PLANNING PHASE

Scenario 1: Important resources are in project area

- *Likelihood:* Moderate (historic industrial garbage found on surface; tribal concern about prehistoric sites in project area)
- *Consequence:* Moderate (costs of resource recovery)
- *Action:* Monitor characterization sampling

Scenario 2: There will be significant delays because of permitting

- *Likelihood:* Low (regulatory approval expected)
- *Consequence:* Low (project delayed)
- *Action:* None

UPON DISCOVERY

Scenario 1: Important resources are in project area

- *Likelihood:* Very High (important resource has been discovered)
- *Consequence:* High (recovery work will be required)
- *Action:* Begin discussions with state archaeology office, historical archaeologists, and local tribes; commission archaeological assessment and recommendations

NOTE

1 *Literature search:* "An examination and review of all written reports (including published, unpublished, reproduced, and manuscript forms), books, articles, etc., pertinent to the investigations carried out for a cultural resource management study. Literature searches differ from records checks in that the latter usually are limited to formalized recorded information which are maintained as reference files" (McGimsey and Davis 1977: 111). *Records check:* "A review of all files containing information relevant to cultural resources. These files may include archeological site survey forms, historic site survey forms (including nomination forms to National and state registers controlled by the State Historic Preservation Officer) photographs, maps, etc. A records check differs from a literature search in that it is more restricted" (McGimsey and Davis 1977: 112).

7

Human Burials

Human burials and cemeteries present a special set of challenges to project managers. As the examples in this book have shown, projects that encounter human remains can have high consequences, especially in parts of the world where local peoples attach great meaning to the remains of their ancestors. Even if local peoples are not emotionally attached to burials in the area, the human remains (bones, associated grave goods, and the surrounding deposits) still must be addressed. Our goal in this chapter is to prepare you for the possibility of uncovering someone's ancestor. We cannot tell you where not to dig, but we can give you a strategy to help minimize the risk of a major discovery of human remains.

The Magnitude of the Problem

Construction projects are continually running into cemeteries, whether during the planning stage, pre-construction stage, or construction stage. The issue is seemingly on everyone's radar screen these days because, quite simply, the impact of encountering burials during project construction can be huge. Costs can be significant, schedule delays disruptive; news coverage gets front-page status, and it can emotionally and physically drain a community.

Communities are continually building on lands that previously have been used to bury the dead. Areas that were once remote are being developed, creating conflicts with people who have lived there for generations. Many of the areas sought after for development are tied to indigenous peoples, who often have a special attachment to the land and the graves of their ancestors. In populated areas the problem is widespread. As cities expand, roads and pipelines get built, new construction projects are funded, the amount of undeveloped land shrinks, and areas once reserved for the dead get developed.

Because of the serious attention afforded human remains in many parts of the world, many countries have passed special laws specific to burials. The United States, for example, passed the Native American Graves Protection and Repatriation Act in 1990, which, among other things, requires a 30-day construction shutdown if Native

American or Hawaiian remains are unearthed during construction on federal lands. Even the international archaeological community has taken steps to establish protocols for its members as part of an effort to stem the increasing destruction of burials and cemeteries (see sidebar, "Guidance from World Archaeological Congress," at the end of this chapter).

WHY ARE HUMAN REMAINS IMPORTANT?

We are frequently asked by project people, "Why are human remains such a big deal?" Some will say, "It's just a bunch of bones."The short answer is: human skeletal remains and burial grounds are considered sacred throughout the world by many people. How that quality of sacredness translates to the way different religions, communities, and countries treat their deceased varies considerably, so this is something you need to check out early in the planning process, whether you are working in Israel, India, or the United States. Recognize, however, that for some groups, especially indigenous peoples, the remains of their ancestors are highly revered. Some groups are willing to fight hard to protect their ancestors, which have come to symbolize their very existence.

In parts of the United States, most Native Americans, for instance, feel there is a strong connection between the living and the dead. Descendants feel it is their responsibility to protect the sacred places where ancestors are located. The buried, they say, are resting and should never be disturbed. If they are moved, harm could occur to those doing the disturbing, as well as to those descendants responsible for protecting their ancestors. While it may be impossible for project personnel to fully understand American Indian concepts about burials, 30 years working in Indian Country have demonstrated to us that the concerns and fears are real.

Further complicating the issue in many areas is that, in addition to bones and grave goods, people attach high spiritual value to the area surrounding the burial ground as well. A sacred place can be important for the memories, the ceremony, the traditional knowledge that it has to offer, or as a place of empowerment. So do not expect that an Indian tribe, indigenous community, or other cultural group will agree to move a burial site just because you have a multimillion dollar project.

ISSUES SURROUNDING MODERN, HISTORIC, AND ANCIENT CEMETERIES

Modern Cemeteries

The nice thing about modern cemeteries is that, for project personnel, they hold few surprises. Since the cemetery is active, its presence is well known. If a project will impact a modern cemetery, those involved in the project should be well aware of its existence, and presumably there is consensus within the community that the cemetery will be moved or that certain measures will be taken to protect it during and after the relocation. These types of removals are not uncommon, especially in areas where land for development is becoming scarce.

Still, removing a cemetery and reburying the people elsewhere is not a simple matter. In the United States, procedures differ from state to state and involve research, detailed planning, and court costs. Generally, the developer must (1) find out who is buried in the cemetery, (2) make a good faith effort to contact the descendants, (3) obtain permission from them to remove the graves to a predetermined location, (4) obtain permission from the city or county authorities to conduct the work, and (5) provide funding. Court costs are inevitable, and dealing with these procedures will require a good deal of time.

Historic Cemeteries

Historic cemeteries are usually more complicated than modern burial grounds. Thousands of family, rural community, and church cemeteries have been forgotten over the centuries, only to be rediscovered by the developer's bulldozer. Usually arrangements can be worked out to everyone's satisfaction, but not always. In the eastern United States, the situation is so common that some companies now specialize in removing historic cemeteries so that development can proceed.

Many times the grave sites have been obliterated and the cemetery boundary can no longer be seen. Small historic cemeteries are especially common in the southern United States, where during the nineteenth century, families buried their family members on their own property. In some places it is estimated there is one family cemetery per square mile. These are frequently found by residential developers, usually during construction. "We didn't know it was on the property when we bought it" (Galbraith 2007) is a comment all too often heard.

Even where grave markers appear on the surface, they probably are not a good indicator of what lies below the surface. They probably don't represent the number of people buried there, because many times people were buried with no headstone, or with a wooden headstone that has long since deteriorated and disappeared. Occasionally, the headstones may have been moved a few feet and are no longer in their original location.

Not knowing that there is a historic cemetery on a land parcel before purchasing it can be expensive. In Georgia (U.S.), for example, a residential developer bought over 100 acres for a housing development. For whatever reasons, he did not investigate the area carefully before purchasing it, and during construction over 20 historic graves were uncovered. He then hired an archaeological firm to help him through the process. The mistake cost him about $40,000 (U.S.). Instead of moving the cemetery, the project proponents had to fence off the cemetery space and incorporate it into the development, further reducing the developer's residential sales (Galbraith 2007).

While moving a historic cemetery can cost as little as $2,000 per grave and legal procedures can be completed in a couple of months, our experiences suggest that costs are generally much higher and legal time much longer than that. Historic American Indian cemeteries can be especially expensive to remove. Most historic cemeteries do not have complete records, and headstones, if they exist, may not match up with the available information.

In one case in the northwestern United States, a historic cemetery had to be moved prior to expansion of a gravel pit. It took three years to collect the documents associated with the people buried there. Local residents said there were only two graves left in the cemetery; the local tribe said there were more. Research, including contacting local funeral homes, cemetery associations, and knowledgeable residents, and reviewing death records, resulted in 13 graves being documented. The cemetery was then hand-excavated by a team of archaeologists and tribal members led by the local tribal cultural resource program. By the time the digging had stopped, 34 Indian and non-Indian graves had been located, exhumed, and relocated. There had been nearly two undocumented burials for each documented burial.

One of the primary factors driving costs is the care with which the burials are removed. It can take from half a day to two days or more to remove a burial and document the vast amount of information contained therein. Descendants may or may not want the burials fully documented, but they certainly will want the remains removed with care. Reburials often occur with a ceremony, which may or may not add further costs. And, in certain cases there may be analytical and reporting costs following the reburial. *Burial Terminology: A Guide for Researchers* (Sprague 2005) provides a good overview of the information that typically needs to be collected when a burial is excavated.

Moving cemeteries is nothing new. For centuries, communities have had to face the problem of competing needs for land. In some instances where historic cemeteries were moved years ago, some graves were inadvertently left behind. This is more likely to have been the case if the cemetery movers used the cemetery fence as their guide. Fences often were constructed years after a place started being used as a cemetery and so do not reflect the true boundary; thus burials are missed. Also, in the past in many places, it was not uncommon for people of other races, people of little means, or community outsiders to be buried outside the fence; this was especially true of church cemeteries, where the unbaptized had to be buried outside the established cemetery.

An example we are familiar with concerns a valuable piece of property that was purchased by a land developer for upscale residential housing near where we live. The developer who had purchased the property was interested in selling the land to another developer. A local land surveyor informed the developer that a cemetery was once located there and that it had been abandoned and relocated nearly 100 years ago. In response, the developer hired an archaeologist to determine whether there would be any obstacles associated with the historic cemetery that would hinder development.

During the archaeological investigation, a Native American burial was discovered. The authorities and local tribe were notified, and the burial was left in place. No other burials were found during the Phase 1 archaeological investigation by backhoe trenching. However, the archaeologist conducted a records search and discovered that there had been a Catholic cemetery located on the knoll above the Native American burial. There was no surface indication that this land had been used as a cemetery; there was no fence, no gravestones, and no depressions.

The historical research indicated that a local Catholic church had started the cemetery in the 1860s. By 1890, the cemetery was becoming overcrowded, so the church moved to a new location and opened a new cemetery. The church sent out a notice to living relatives to come and exhume their deceased family members. Early records stated that most of the graves were relocated, and statements by the local Catholic church stated that all of the bodies formerly interred at that location had been exhumed and relocated.

Due to the likelihood that burials still remained in the area, the developer decided to pay for archaeological investigations to find the original boundaries and any additional remains. During the archaeological investigation, the boundaries were determined and many historic graves were located. A potter's field, an area where "undesirables" were buried, was also located on the hill adjacent to the cemetery. Due to the huge financial burden of removing and relocating the remaining graves, the developer left the cemetery intact. Today the knoll is well vegetated and undeveloped. The developer paid for the archaeological study but lost the sale of the property; however, he also eliminated the risk of an expensive burial removal project. The developer had been well served by his archaeological consultant.

Landownership and management responsibilities can loom large in resolving problems surrounding historic cemeteries. Many historic cemeteries are now on private land, owned by people with no historical ties to those buried in the cemetery. Others are owned by cemetery associations that are now defunct, and ownership is unclear. And sometimes people lay claim to abandoned cemeteries that are situated on land to which they have no legal right. When you combine ownership issues with governmental laws passed to protect historic cemeteries, relocating a historic cemetery may be possible, but it is unlikely to be easy or inexpensive.

Ancient Burial Grounds

For the purpose of this book, prehistoric burials or ancient burials are those that date before the written record, before European contact in the United States, and this time frame varies depending on where you are geographically. In the United States, Indian people were buried where they died or in a burial ground where others of the community were buried. Through time, natural factors, such as wind and rain, may have caused erosion which may have brought the burials out onto the surface or, conversely, more deeply buried into the hillside. Descendant populations can be hard to track down, given the extensive movements of tribes across the continent in the last few hundred years. Unless a particular group of people claim the individual or burial ground, identification or cultural affiliation may be lost.

Prehistoric burials are problematic because they are almost never marked, at least not in the way that modern people can readily recognize. These are the burials that are most likely to be found inadvertently during construction. Small developers seem to run into burials more often than large developers, because they are less likely to fully research an area before they purchase it. Large developers usually have an archaeologist on staff or have the money to pay for this expertise and many times can avoid encountering unexpected burials.

In the United States, American Indian burials raise complex issues that are not easily solved. There are thriving Indian communities whose ancestors are indigenous to this land. They have relationships with the land, and their ancestors have been buried there for thousands of years. Today the indigenous peoples around the world have the power and control over their burials that they once lacked. There are now laws that protect these cemeteries, and most indigenous peoples know their rights. They have a voice at the table and have shown that they will use it. A common goal is the return, sometimes referred to as *repatriation*, of the remains to the descendants.

Occasionally, repatriation of remains to one group can fuel additional controversy—for example, if more than one group claim to be the descendants and responsible for the remains. Politics can enter the equation when land disputes are involved; that is, one group may fear it is relinquishing legal claims to an area if it allows some other group to take the remains. This is not a common problem, but it can happen, and it can impact project schedules if burial recovery is slowed while negotiations continue among agencies and groups.

Similarly, and more common in Western countries, are controversies among the cultural groups laying claim to the remains. Some do not want scientific study and intend to rebury the remains when they get them. Others, who view the skeletons as scientific specimens, want to conduct destructive and nondestructive studies (Killion 2008). These situations tend to be controversial and can be time-consuming and costly if not planned for in advance. The Kennewick Man case from Washington State, where local tribes and scientists fought over control of the remains for years, is a good example of the costs and delays that can occur when the courts become involved (Burke et al. 2008). We will not address this debate further here, except to say that the issue of whether or not to scientifically study the remains is one that is hot and emotionally packed.

If the project is located in an area where archaeologists are the only interested party, the developer may be required to pay for scientific study of human remains as part of the documentation and reporting process. The project may also be required to allow time and funds to repatriate and rebury the remains. There may be state funds or other resources available to assist with these costs, but such funds are usually scarce. You will need to check the laws that apply to the project area prior to construction.

Rather than view indigenous groups as adversaries, project managers need to seek them out and involve them in the process, even if the project is not required to do so. Many indigenous groups these days have cultural programs that hire archaeologists; these programs can be great sources of historic information. Even if there is no program, indigenous group members can help with your background research of the project area and will usually share their concerns and needs. They have traditional knowledge that is passed down through time and can help you identify the potential for burials. They may not always reveal the details, but they will teach you what you need to know, if you are willing to listen.

How to Avoid Burial Grounds

To avoid unearthing a major burial ground, do your homework before buying the land or choosing your project location. It is imperative that the project take a realistic approach to the potential for a human remains discovery from the moment a project begins. As an American Indian friend of ours once said, "If you don't do research, you're rollin' dice." You need to know the history of the project area from recent times to as far back as you can go. Every major archaeological disaster associated with cemeteries could have been prevented if the project manager had either commissioned an initial cultural assessment or had listened to people who expressed concern about human remains in the area.

If there is any possibility that the proposed project area contains a cemetery, efforts need to be made to understand how the community members feel about their ancestral cemeteries. How might a living relative in the area feel if their loved one or ancestors were uncovered during the project. If you don't know how the community might feel, just ask.

Avoiding a major cemetery discovery requires a change in mind-set from the past. Developers have long feared the discovery of human bones. They fear that the bones will shut down the project, delay it for an extended period of time, and cost endless amounts of money for relocation. This fear has led people to avoid the issue, pretend it did not exist, play the odds. Today, it simply is not worth the risk to take that approach.

A final tip: Beware of the "accelerated" project. There has been an uncanny coincidence for large, significant, costly discoveries to be made in projects that were labeled, "accelerated," "streamlined," or "fast-tracked." When such cases are examined in retrospect, we find that "accelerated" somehow got translated into cutting out the upfront research, ignoring basic sampling principles, ignoring review processes, and ignoring those who express concerns.

WHAT TO DO IF BURIALS ARE DISCOVERED

It should be part of every land developer's pre-business plan to ensure that there are no burials in the project area before the purchase is finalized. Had the Washington Department of Transportation done this before buying the 22.5 acres from the City of Port Angeles, it would have saved the taxpayers over $80 million. Similarly, it is critical that the initial assessment discussed in Chapter 3 explicitly target burials.

Suppose you have done everything you can to ensure there are no burials to be discovered. Then, three days into the excavation your project uncovers a significant number of bones. What do you do? Assuming you have prepared an inadvertent discovery plan, you will have procedures already in place so your response will be clear. You will have already met and consulted with the relevant indigenous groups or members of the local community. The interested parties will already be on board, and recovery time should be relatively short. There will be some extra costs involved, but you should have budgeted for this possibility in your contingency budget.

Upon discovery of bones, it is important to determine as soon as possible whether the bones are human or animal. If a human skull has been unearthed, then there is no question in anyone's mind whether the remains are human or not. More commonly, the bones found are fragmentary and are therefore more difficult to identify. Ideally, the person making the determination should be trained in human bone identification, but people with these qualifications are not always available. If the project team includes someone like an environmental scientist, or if there is a member of the community who has training as a veterinarian, nurse, dentist, doctor, or x-ray technician, that person might be able to make a determination using a human bone identification field manual (e.g., White and Folkens 2005).

If, at any time, you think you have unearthed human bones, protocol in most places is to call the police, as it might be a crime scene. This is important. Law enforcement and other governmental contacts should be identified in the project's inadvertent discovery plan. Be sure to protect the site from further disturbance until directed otherwise. It is important to cause as little disturbance as possible so that no information is destroyed.

If the bones are not human, you will need to determine whether you have uncovered an archaeological site. Contact the archaeological expert identified in the inadvertent discovery plan to make the determination. If possible, do not pick up the bones, as leaving them in place will make it easier for a professional to make a determination.

If human bones are identified, be respectful and follow the law. The following list provides actions that typically are followed:

1. Shut down work in the immediate area. This does not mean you have to shut down the entire project; it means you cannot dig in the immediate area until the situation has been resolved. Move workers away from the area and flag it off.
2. Contact the authorities. Do not remove or alter anything at the site. Do not reenter this area, as it could be a crime scene.
3. Secure the area. While waiting for authorities, secure the area with "caution tape" or some sort of barrier for protection. If the area is heavily traveled, request extra security from local law enforcement.
4. Recheck regulations. Increasingly, states and local governments have procedures on how burials should be handled, who should be contacted, and what reports need to be filed.
5. Inform agencies, tribes, and governments. Necessary contacts may be made informally, or protocol may require formal measures such as letters or meetings.
6. Proceed as directed by authorities. If the discovery is a crime scene, then authorities will begin law enforcement procedures. If the discovery is old, said to have antiquity, or is ancient, or archaeological, then the governmental heritage professionals can help you locate the indigenous or community group that would be responsible for the remains. In the United States, contact the State Historic Preservation Office (SHPO) and report the findings. They will assist you in notifying the appropriate Native American tribe and a local archaeologist who has expertise in the locale.
7. Consult with the responsible cultural group. If a group claims the finds, they will proceed according to the law and their beliefs. Help them.
8. Be prepared to get out of the way. Let others come in and do their job.

Many times relocating the cemetery is the best option. In Israel, Gaza settlers were disinterred from the Gaza Strip and reburied elsewhere (Engber 2005). In the United States, cemeteries are moved for highways, shopping malls, and housing. The procedures for relocation vary from country to country and, in the U.S., from state to state. Relocating is not the only option, however. Many times it is easier and less costly to leave the cemetery in place and build around it. Either way, your project will require early planning and some careful historical research to determine if bones, burials, or cemeteries exist in your right-of-way.

In summary, when an inadvertent discovery of human remains is encountered, avoid walking over, removing, or further disturbing the burial. Any artifacts should be left untouched and unmoved from the location of discovery. Avoid taking photographs of the burial unless there is an emergency situation or you are directed to do so by authorities. The media may show up with the coroner. Try to reschedule an interview after authorities and responsible groups have had time to respond. Always be respectful.

We close this chapter with a short essay by our colleague Mary Rossi from the Pacific Northwest. Mary provides a first-hand account of how showing respect for human remains is not only the right thing to do, but can accommodate your project.

RESPECT FOR HUMAN REMAINS: SOUND RISK MANAGEMENT FOR PROJECTS AND FOR RELATIONSHIPS

by Mary Rossi

For over 11 years, I have lived, studied, and worked in Washington state where recent archaeological disasters have all too frequently involved human remains. Only eight months out of graduate school, I was hired by the Lummi Nation to assist with the recovery of thousands of fragmented human remains and associated artifacts unearthed during the expansion of a sewage treatment plant in the City of Blaine. Three years in the field and three years as the Lummi Nation's first Tribal Historic Preservation Officer (THPO) taught me that such disasters have devastating impacts on everyone involved—tribal communities, local communities, government agencies, cultural resource professionals, developers, general contractors, politicians, educators, students, and the general public who witnesses and sometimes pays the expenses. As I began to understand the potential risks to human remains associated with development projects, I decided to focus my efforts on doing whatever I could to prevent such disasters, primarily through active and (hopefully) effective involvement in the development process.

As I came to realize the risks associated with development projects, I also saw that we all have an opportunity to manage risks through early and effective project planning. Not only do we have the opportunity to plan, we also have an obligation to do so, because the failure to plan carries great risks both to projects and to our relationships with those affected by our actions. Respect for human remains and the graves and cemeteries where they are found is commonly accepted as a fundamental value of nearly every culture, so we must work to design and implement projects in a way that ensures such respect.

As a self-described "cultural resource planner," I think a lot about how my chosen profession can contribute in a positive manner to better project planning. Perhaps one of the most important things we must first do is to look up from our own, often all-consuming work and identify our partners. In other words, who should we be educating about what we are trying to do, and who will be affected by our actions? Since an ever-increasing amount of cultural resource work, particularly archaeology, is driven by development, cultural resource professionals benefit from working closely with partners such as real estate developers and construction personnel.

If the first step to better project planning is to recognize the partners, the next step is to determine how to work with them in a productive manner. Initially, we might ask what the partners need to know about human remains; the exchange will be more instructive if it is a two-way conversation, so we might also ask what the cultural resource profession needs to know about development. For better or worse, my educational experience did not include much, if any, business training, so I have had to commit to learning new business skills and processes, not to mention new terminology and methods of problem solving. Ideally, this approach to identifying and building partnerships will allow me to educate and assist developers more effectively, thereby resulting in an increased awareness of human remains and the people who care about them.

If developers and the construction industry are important partners, how do cultural resource professionals help them manage risk through better planning? One way to begin managing risk is through education about the points at which the development process and human remains intersect. One such intersection is the environmental permitting process. Cultural resource professionals should have a working knowledge of what permits are needed when, and they should communicate this to developers so they are able to incorporate the permitting processes into broader, multivariant project schedules. Human remains represent only one of a growing number of development considerations, so professionals must know where they fit into the larger picture. Respectful protection of human remains depends on our ability to communicate across professions.

Related to the permitting process are the government agencies responsible for reviewing permit applications and issuing the permits themselves. Cultural resource professionals must understand and then educate developers about agency procedures, requirements, and timelines. Education leads to effective execution, which, in turn, leads to significant savings of time and money, something that developers understand and appreciate and something that benefits all parties in the long run.

Developers and construction personnel will also benefit from understanding how cultural resource professionals help them realize their goals. Services such as identification, assessment, and management will keep developers in compliance with applicable regulations and, if successfully executed, will prevent the unnecessary destruction of human remains. Partnerships based on education and trust serve as the foundation for more effective management of risks to both projects and human relationships.

Finally, building upon the types of partnerships described, cultural resource professionals can advocate for earlier planning efforts so that options are discussed even before permit applications are submitted and permit clocks start ticking. Developers regularly employ planning methods such as fatal-flaw analyses and feasibility studies, but are cultural resources considered during these phases of development? If not, they should be added to the list of considerations. Earlier identification and assessment affords more time for the discussion of viable design options should project adjustments prove necessary. Conversely, discoveries occurring after the construction process has begun leave little or no time for comprehensive problem solving. In many states, including Washington, legislation mandating work stoppages and consultation is being passed in response to the unpredictable nature of so-called inadvertent discoveries. Perhaps if we all do a better job of pre-project planning, such discoveries will become a thing of the past.

If we are successful in our planning efforts, human remains will receive the respect they deserve, as will the communities caring for them. Projects and relationships based on understanding and trust will become the norm. Instead of confusion and trepidation, the cultural resource profession will instill confidence and certainty in its partners. Is this a lofty goal? Perhaps. Is it necessary? Absolutely. The remains of our ancestors depend on it.

GUIDANCE FROM THE WORLD ARCHAEOLOGICAL CONGRESS

THE VERMILLION ACCORD ON HUMAN REMAINS ADOPTED IN 1989 AT WAC INTER-CONGRESS, SOUTH DAKOTA, USA

1. Respect for the mortal remains of the dead shall be accorded to all, irrespective of origin, race, religion, nationality, custom and tradition.
2. Respect for the wishes of the dead concerning disposition shall be accorded whenever possible, reasonable and lawful, when they are known or can be reasonably inferred.
3. Respect for the wishes of the local community and of relatives or guardians of the dead shall be accorded whenever possible, reasonable and lawful.
4. Respect for the scientific research value of skeletal, mummified and other human remains (including fossil hominids) shall be accorded when such value is demonstrated to exist.
5. Agreement on the disposition of fossil, skeletal, mummified and other remains shall be reached by negotiation on the basis of mutual respect for the legitimate concerns of communities for the proper disposition of their ancestors, as well as the legitimate concerns of science and education.
6. The express recognition that the concerns of various ethnic groups, as well as those of science, are legitimate and to be respected will permit acceptable agreements to be reached and honored.

8 Principles for Avoiding an Archaeological Disaster

In this book we have argued that heritage resources, primarily archaeological resources, can cause serious problems for large-scale construction and development projects. We have provided a framework to help project managers evaluate when they should take action to keep heritage resource issues from causing serious disruptions to project schedules and budgets. In this closing chapter, we offer the following heritage management principles for project managers and project staff to follow as they navigate their way through a sensitive project.

Our ten principles for avoiding an archaeological disaster are:

1. *Actively manage the heritage resource component of the project* just as you manage other critical project components.
2. *Hire professional, qualified expertise* to advise you on heritage resource issues and to conduct the assessment and fieldwork needed.
3. *Learn the regulatory requirements* that have been established for your project area.
4. *Identify and consult* with interested parties regarding the heritage resources important to them.
5. *Conduct a comprehensive background research analysis* and site records check on the project area and don't fall for the "it's disturbed" claim unless your research can confirm the disturbance.
6. *Incorporate the archaeological risk management approach* into project decision making.
7. *Minimize destruction of heritage resources wherever possible* because they are non-renewable and important to communities.
8. *Be prepared for the unexpected; have contingency plans in place.*
9. *Be open, transparent, and honest from the beginning.*
10. *Just follow the process* and don't fall prey to accelerated approaches, streamlining, or other innovative approaches if they violate basic heritage management principles.

CONCLUDING COMMENT

We hope that our book has demonstrated the value that the archaeological risk management approach can have for your project. If we have increased your awareness of heritage resources and the basic actions you can take to protect your project from inadvertently destroying important resources, then we are well on our way to meeting our goal of "no more archaeological disasters."

AFTERWORD: Sacramento City Hall Expansion Project: Another Inconvenient Truth

by Adrian Praetzellis

No one should have been surprised when construction workers uncovered a Native American cemetery during excavations for Sacramento's new city hall garage in April 2003. After all, it was only a couple hundred feet from a known prehistoric site, and an archaeologist had told city officials what to expect months before construction began.

The City Hall Expansion project was a 267,000 square foot office and parking structure to be built immediately to the north of the historic City Hall (City of Sacramento 2003:1). Immediately to the south, however, was CA-SAC-30, a prehistoric site recorded decades before and known to contain human remains. In 2001 the city issued the project's environmental impact report, which contained the usual admonition that construction must be temporarily halted if human remains were found during construction. Later that year, city officials met with an archaeologist, explained the construction project, and asked for recommendations. For the archaeologist, it was a no-brainer: this was a perfect case for pre-construction testing—to use drilling equipment and an excavator to punch through the deep layer of historic fill to see if CA-SAC-30 extended into the construction zone. If the prehistoric site was there, the construction schedule could be adjusted to enable archaeological recovery and the project could move ahead.

Deciding that construction monitoring by an archaeologist was a better plan, the city set into motion a course of events that resulted in unnecessary cost and delay, for shortly after construction began, human remains began to turn up. Attorneys for Jeff Murray, Vice-Chair of the Shingle Springs Band of Miwok Indians, immediately sought a restraining order against the city to halt the project (*Sacramento Bee* 2003a, 2003b, 2003c). The city and Murray's group eventually came to an agreement, and an archaeologist was hired to excavate the site, which produced approximately 75,000 artifacts and 45 sets of human remains dating ca. 4,500 years before the present (Sacramento County 2006: 3–36).

By the time construction resumed nearly three months later, the discovery had cost $4 million, necessitating that the City issue new construction bonds to pay for the now $85 million project (KRCA 2003). Weeks of schedule slip, nearly $2 million in delay-related costs, and a public relations nightmare would have been avoided if City officials had taken the warning signs seriously.

Appendix A
World Bank Guidance on Issues Surrounding Cultural Property

1. "Cultural property" refers to sites, structures, and remains of archaeological, historical, religious, cultural, or aesthetic value. Many projects or project components have a potential impact on archaeological sites, building complexes, architecture, monumental sculpture, painting, inscriptions, and other physical remains left by previous human inhabitants and considered part of a country's cultural heritage. In numerous cases cultural property sites coincide with important natural sites. Cultural resources are a part of the resource base, and it is therefore important that development options under consideration are screened for potential impacts on cultural property. As discussed in the World Bank Operational Directive 4.50: "Cultural Property" (Draft), the Bank's approach to the conservation and management of cultural property is to assist in the protection and enhancement of cultural property affected by Bank-financed projects.

Policy, Procedures, and Guidelines

2. The Bank recognizes that socially stable development requires societies to retain and keep alive ties to their past and their cultural traditions. The Bank's policy as stated in Operational Directive 4.50 (which updates Operational Policy Note 11.03: "Management of Cultural Property in Bank-Financed Projects") is to assist in protecting and enhancing cultural property through specific project components and decline to finance projects which significantly damage cultural property, and assist only those that are designed to prevent or minimize such damage. Bank Operational Directives on environmental assessment, involuntary resettlement, and tribal peoples also deal with cultural property concerns.

Relationship to Bank Investments

3. Recognizing the value of their cultural heritage, most countries have enacted legislation aimed at the protection of their cultural properties. Unless development projects address the interests of archaeological and historical conservation as an integral element in planning, expensive delays can result and cultural resources can be inadvertently destroyed. Contractors can avoid costly construction delays by maintaining close contact with the appropriate department of antiquities or national museum of archeology. If archaeologists are notified well in advance, they can complete their investigation of a

site before the construction begins, without affecting the contractor's schedule. If antiquities are encountered in the course of construction, a team from the department of antiquities and/or an international archaeological institute usually needs only a few days to excavate and record vital information. In most cases, the construction work can continue while the archaeologists are on site.

4. Any project which involves excavation, leveling or filling of earth as part of construction operational practices is a potential threat to archaeological and historical remains (see OD 4.01, Annex A: "Environmental Assessment" and Chapter 1 of the Sourcebook for a listing of environmental screening category A and B projects). While it is not possible to generalize about the extent of impact of projects on cultural property, infrastructure and energy are the sectors most affected by cultural resource issues. Both urban and rural development schemes may also encounter cultural property in carrying out urban infrastructure improvement: road building, land reclamation, irrigation, drainage, etc. Public facilities (schools, hospitals, housing, etc.) can also affect cultural property if facilities are constructed upon or adjacent to archaeological sites.
5. The relationships between cultural property issues and a project can range from direct to indirect. A hydropower project may result in the inundation of cultural sites unless mitigating measures are implemented or relocation of the project is undertaken. A new road adjacent to an archaeological site or historic building may facilitate access and therefore increase the cultural property's vulnerability. Construction work near archaeological or historical remains may cause damaging vibrations and disturbances. The location of a new industrial facility may bring cultural resources into contact with airborne pollutants. Environmental projects aimed at improving air quality may have a notable effect on decreasing air pollution that damages stone structures. Reducing drainage problems associated with both irrigation and wastewater, which disturb building foundations, may also have a direct favorable impact on the condition of historic structures.

Guidance for Environmental Assessments

6. At the earliest stages of project identification and preparation, it is incumbent on the task manager, with advice and operational support provided by the Regional Environment Division (RED) and the Environment Department (ENV), to alert governments to potential cultural property issues. A first step would be to check the list of UNESCO World Heritage Sites, which now numbers over 300 sites defined as constituting a recognized heritage of outstanding universal value. Responsible government agencies (including ministries of antiquities or culture), museums, university departments of archaeology, art history, or architecture, should be contacted for information and expertise about the cultural resources in the project area. National inventories of cultural resources can provide important data to assist project officers. International research institutes (foreign archaeological missions are present in a number of countries) are also sources of expertise. Relevant nongovernmental organizations, such as the national committees of the International Committee of Monuments and Sites (ICOMOS), and the

International Centre for the Study of the Preservation and the Restoration of Cultural Property (ICCROM), may also be helpful in gaining a full understanding of the cultural property that may be affected, its relative importance, and conservation needs. International organizations with expertise in cultural property protection (for example, UNESCO) are other sources of expert advice.

7. To coordinate the various government agencies, a representative of the department of antiquities should be invited to participate in interagency meetings held at key points in the EA cycle, as well as in forum meetings with representatives of affected groups and relevant non-governmental organizations (NGOs).
8. For certain types of projects that involve large-scale modification and/or disturbance of earth surfaces (such as dams, irrigation systems, ports, principal roads, pipelines, subways, sewers, mines, and land reclamation) and that are located in areas where there is reasonable evidence to believe cultural resources are present, a field survey by a qualified expert (for example, an archaeologist or architectural historian) is required. Informed decisions can be made about subsequent actions on the basis of this survey. Full inventories and further studies may be needed.
9. Detailed analyses, including review of project alternatives, mitigation measures, and institutional, training and monitoring requirements, should be used to assist the project executing agency and Bank staff in deciding whether to redesign or relocate the project so that sites and structures can be preserved, studied, or conserved intact in situ. In certain cases, it may be feasible for structures and remains of cultural value to be relocated, restored, preserved, and studied on alternative sites. Archaeological field survey, selective salvage, and preservation of excavated finds in museums prior to large-scale excavation, land leveling or tilling may help to minimize the loss of or damage to cultural property.
10. Sites that are buried or not located by the survey may be discovered during project implementation, especially in the course of construction or mining. Such unanticipated discoveries of remains of an archaeological and/or historical nature, termed archaeological chance finds, are frequently found within 0–3 meters of the present surface. Most often they are concentrations of pottery, worked stone, and human and animal bones, without commercial value, but of significance to archaeologists, historians, anthropologists and paleontologists. In general, the following archaeological chance find procedures should be adopted in project design and construction contracts: notification of the relevant department of antiquities; request for a representative to make a site inspection; cessation of work in the vicinity of the find until the visit of a representative; and decision by department of antiquities on possible salvage or excavation (usually required within 48–72 hours of notification). National laws normally assess penalties for violations of archaeological chance find procedures.
11. In projects with major impacts on cultural resources, consideration should be given to setting up a conservation unit located in a ministry or line agency. Efforts should also be made to develop local institutional capacity through technical assistance and training.

12. Archaeological and historic sites, settlements, and structures are vulnerable to disturbances of various types. Among these are flooding, changes in water table, deterioration of air quality, coastal erosion, seismic disturbances, vibrations caused by construction activities, building, or airplanes, and human intrusion. Cultural sites can tolerate finite numbers of visitors, just as natural sites, and this should be assessed in project design. The number of visitors and areas of access need to be controlled in order to prevent sites from deterioration due to overuse and physical proximity (visitors touching walls, paintings, sculptures).
13. It is not enough to acknowledge the existence of cultural property that will be affected by a Bank project. The site's long-term sustainability should be considered and a well designed management plan be devised that addresses conservation and maintenance requirements. In this context, the Bank is willing to provide technical assistance for training in conservation and management.
14. Where sites are considered sacred, as in the case of religious shrines, the impact of relocation is complex. In such instances it is strongly advised that a team be formed to develop mitigation measures. The team should have an art or architectural historian knowledgeable about the particular cultural tradition, an architectural conservator, an anthropologist familiar with the population of the area, and a coordinator who would bring together the relevant government organizations, experts, and community leaders. It is important that such interventions be scientifically sound, and that they respond, as completely as possible, to patterns of social organization and existing social and cultural institutions. Project staff should ensure that the cultural heritage of non-dominant cultures are accorded the same care as that of the dominant cultures.

Source: World Bank. 1999. *Environmental Assessment Sourcebook*. Chapter 3. Social and Cultural Issues in Environmental Review. Permanent URL for this page: http://go.worldbank.org/CR3COCT8P0. Accessed November 16, 2008.

Appendix B
Professional Archaeological Organizations

American Cultural Resources Association (ACRA)

ACRA is a nonprofit trade association that supports the business needs of the diverse cultural resource management industry. Members represent all aspects of the cultural resource industry, including historic preservation, history, archaeology, architectural history, historical architecture, landscape architecture, and specialty subfields such as geoarchaeology, soil science, and ethnobotany.

ACRA's mission is to:

- Promote professionalism, ethics, and best practices throughout the industry;
- Influence federal and state preservation policy for the benefit of cultural resources;
- Provide ongoing professional training and educational opportunities for ACRA members;
- Promote public awareness of the cultural resources management and all of its affiliated disciplines;
- Assist our members in developing, promoting and improving their businesses.

Website: *http://acra-crm.org*

Archaeological Institute of America (AIA)

The Archaeological Institute of America promotes a vivid and informed public interest in the cultures and civilizations of the past, supports archaeological research, fosters the sound professional practice of archaeology, advocates the preservation of the world's archaeological heritage, and represents the discipline in the wider world.

Website: *www.archaeological.org*

Australian Association of Consulting Archaeologists (AACAI)

The Australian Association of Consulting Archaeologists Inc. (AACAI) is an organization for professionals working in all fields of contract and public archaeology. It aims to uphold and promote the discipline and to advance the welfare of members. AACAI has a Constitution, a Code of Ethics and a Consulting with Aboriginal Communities Policy Document. It is affiliated with the Australian Archaeological Association Inc and is a Foundation Member of the Council for the Humanities, Arts and Social Sciences.

Website: *www.aacai.com.au*

CANADIAN ARCHAEOLOGICAL ASSOCIATION (CAA)

The Canadian Archaeological Association (CAA) was founded in 1968. Membership includes professional, avocational, and student archaeologists, as well as individuals of the general public of any country, who are interested in furthering the objectives of the Association.

The objectives of the CAA are as follows:

- To promote the increase and the dissemination of archaeological knowledge in Canada;
- To promote active discourse and cooperation among archaeological societies and agencies and encourage archaeological research and conservation efforts;
- To foster cooperative endeavors with aboriginal groups and agencies concerned with First Peoples' heritage of Canada;
- To serve as the national association capable of promoting activities advantageous to archaeology and discouraging activities detrimental to archaeology;
- To publish archaeological literature;
- To stimulate the interest of the general public in archaeology.

Website: *www.canadianarchaeology.com*

EGYPTIAN CULTURAL HERITAGE ASSOCIATION (ECHO)

The Egyptian Cultural Heritage Association is a charitable organisation committed to assisting young Egyptians in preserving and managing the world heritage of Egypt as the most effective, long-term, sustainable, and least expensive means of safeguarding the threatened archaeological record.

Website: *www.e-c-h-o.org/*

EUROPEAN ASSOCIATION OF ARCHAEOLOGISTS (EAA)

The European Association of Archaeologists is the association for all professional archaeologists of Europe and beyond. The EAA is a membership-based association open to all archaeologists and other related or interested individuals or bodies. The EAA currently has over 1,100 members on its database from 41 countries worldwide working in prehistory, classical, medieval, and later archaeology. They include academics, aerial archaeologists, environmental archaeologists, field archaeologists, heritage managers, historians, museum curators, researchers, scientists, teachers, conservators, underwater archaeologists, and students of archaeology.

Website: *www.e-a-a.org*

REGISTER OF PROFESSIONAL ARCHAEOLOGISTS (RPA)

The Register of Professional Archaeologists is a listing of archaeologists, primarily from North America, who have agreed to abide by an explicit code of conduct and standards of research performance, who hold a graduate degree in archaeology, anthropology, art history, classics, history, or another germane discipline and who have substantial practical experience. Registration is a voluntary act that recognizes an individual's personal responsibility to be held accountable for their professional behavior. By formally acknowledging this relationship between personal actions and the wider discipline of archaeology, the act of registration sets the professional archaeologist apart from all others who are involved with or interested in archaeology.

Website: *www.rpanet.org*

SOCIETY OF AFRICANIST ARCHAEOLOGISTS (SAfA)

The Society of Africanist Archaeologists is an organization of archaeologists, researchers from associated disciplines, and others who share an interest in African archaeology and African societies. Membership is international, with participation from Africa, the Americas, Europe, and Asia, and is actively involved in research in many African countries. SAfA is an affiliated organization of the Society for American Archaeology and the African Studies Association.

Website: *http://cohesion.rice.edu*

SOCIETY FOR AMERICAN ARCHAEOLOGY (SAA)

The Society for American Archaeology (SAA) is an international organization dedicated to the research, interpretation, and protection of the archaeological heritage of the Americas. With more than 7,000 members, the society represents professional, student, and avocational archaeologists working in a variety of settings, including government agencies, colleges and universities, museums, and the private sector.

Since its inception in 1934, SAA has endeavored to stimulate interest and research in American archaeology; advocate and aid in the conservation of archaeological resources; encourage public access to and appreciation of archaeology; oppose all looting of sites and the purchase and sale of looted archaeological materials; and serve as a bond among those interested in the archaeology of the Americas.

Website: *www.saa.org*

SOCIETY FOR HISTORICAL ARCHAEOLOGY (SHA)

Formed in 1967, the Society for Historical Archaeology (SHA) is the largest scholarly group concerned with the archaeology of the modern world (A.D. 1400–present). The main focus of the society is the era since the beginning of European exploration. SHA promotes scholarly research and the dissemination of knowledge concerning historical archaeology. The society is specifically concerned with the identification, excavation, interpretation, and conservation of sites and materials on land and underwater. Geographically the society emphasizes the New World, but also includes European exploration and settlement in Africa, Asia, and Oceania.

Website: *www.SHA.org*

World Archaeological Congress (WAC)

The World Archaeological Congress (WAC) is a nongovernmental, not-for-profit organization and is the only representative worldwide body of practicing archaeologists. WAC seeks to promote interest in the past in all countries, to encourage the development of regionally based histories, and to foster international academic interaction. Its aims are based on the need to recognize the historical and social roles as well as the political context of archaeology, and the need to make archaeological studies relevant to the wider community.

WAC is an international forum for discussion among individuals concerned with the study of the past. WAC holds an international congress every four years to promote the exchange of results from archaeological research; professional training and public education for disadvantaged nations, groups, and communities; the empowerment and support of indigenous groups and First Nations peoples; and the conservation of archaeological sites.

Website: *www.worldarchaeologicalcongress.org*

Listings of many other national and international societies can be found on the About.Com: Archaeology website: *http://archaeology.about.com/od/smokepfeiffer/Michael_A_Smoke_Pfeiffers_Directory_of_Archaeology.htm.*

Appendix C
Selected Ethical Codes from Heritage Management Associations

This appendix presents a selection of ethical codes supported by a variety of heritage resource professional associations. We have chosen those codes that relate most directly to development and construction projects. The complete codes for each organization can be found at their respected websites.

The American Cultural Resources Association (ACRA)

The ACRA Member's Responsibilities to Clients

An ACRA member is obligated to provide diligent, creative, honest, and competent services and professional advice to its clients. Such performance must be consistent with the ACRA member's responsibilities to the public interest. An ACRA member shall:

- Exercise independent professional judgment on behalf of clients.
- Accept the decisions of a client concerning the objectives and nature of the professional services provided unless the decisions involve conduct that is illegal or inconsistent with the ACRA member's obligations to the public interest.
- Fulfill the spirit, as well as the letter, of contractual agreements.
- Not provide professional services if there is an actual, apparent, or perceived conflict of interest, or an appearance of impropriety, without full written disclosure and agreement by all concerned parties.
- Not disclose information gained from the provision of professional services for private benefit without prior client approval.
- Not solicit prospective clients through the use of false or misleading claims.
- Not sell or offer to sell services by stating or implying an ability to influence decisions by improper means.
- Not solicit or provide services beyond the level or breadth of the professional competence of its staff or project team.
- Solicit or provide services only if they can responsibly be performed with the timeliness required by its clients.

- Not solicit or accept improper compensation for the provision of judgments or recommendations favorable to its clients.
- Not offer or provide improper compensation as a material consideration in obtaining or sustaining client or prospective client favor.
- Disclose information identified as confidential by its client only if required by law, required to prevent violation of the law, or required to prevent injury to the public interest.

Website: *http://acra-crm.org/displaycommon.cfm?an=1&subarticlenbr=20*

ARCHAEOLOGICAL INSTITUTE OF AMERICA (AIA)

The Responsibilities to the Archaeological Record

Professional archaeologists incur responsibilities to the archaeological record —to the physical remains, including those located under water, and to all the associated information about those remains.

1. The purposes and consequences of all archaeological research should be carefully considered before the beginning of work. Approaches and methods should be chosen that require a minimum of damage to the archaeological record. Although excavation is sometimes the appropriate means of research, archaeological survey, study of previously excavated material, and other means should be considered before resort is made to excavation.
2. The recovery and study of archaeological material from all periods should be carried out only under the supervision of qualified personnel.
3. Archaeologists should anticipate and provide for adequate and accessible long-term storage and curatorial facilities for all archaeological materials, records, and archives, including machine-readable data, which require specialized archival care and maintenance.
4. Archaeologists should make public the results of their research in a timely fashion, making evidence available to others if publication is not accomplished within a reasonable time.
5. All research projects should contain specific plans for conservation, preservation, and publication from the very outset, and funds should be secured for such purposes.

Website: *www.archaeological.org/pdfs/AIA_Code_of_Professional_StandardsA5S.pdf*

CANADIAN ARCHAEOLOGICAL ASSOCIATION (CAA)

Professional Responsibilities

Archaeological remains are finite, fragile, non-renewable and unique. Before undertaking responsibility for any excavation that destroys a portion of the archaeological record, members of the Canadian Archaeological Association must:

- Keep abreast of developments in their specializations;
- Possess adequate training, support, resources and facilities to undertake excavation and analysis;

- Produce an adequate document worthy of the destruction of the archaeological remains;
- Present archaeology and research results in a timely and responsible manner;
- Preserve documentation in such a way that it is of value to future researchers;
- Comply with all legislation and local protocols with Aboriginal Peoples, as described in the Statement of Principles for Ethical Conduct Pertaining to Aboriginal Peoples, as appropriate in each province and/or territory;
- Respect colleagues, and cooperate with them;
- Allow the expression of alternative views of the past;
- Exercise the right to defend our own scholarship;
- Recognize that documentation of an archaeological record should, within a reasonable period of time, become available to others with legitimate research interests;
- Present archaeological information in an objective and well informed manner in all contexts.

Website: *www.canadianarchaeology.com/conduct.lasso*

REGISTER OF PROFESSIONAL ARCHAEOLOGISTS (RPA)

Codes of Conduct

The Archaeologist's Responsibility to Employers and Clients:

- An archaeologist shall:
  a. Respect the interests of her/his employer or client, so far as is consistent with the public welfare and this Code and Standards;
  b. Refuse to comply with any request or demand of an employer or client which conflicts with the Code and Standards;
  c. Recommend to employers or clients the employment of other archaeologists or other expert consultants upon encountering archaeological problems beyond her/his own competence;
  d. Exercise reasonable care to prevent her/his employees, colleagues, associates and others whose services are utilized by her/him from revealing or using confidential information. Confidential information means information of a non-archaeological nature gained in the course of employment which the employer or client has requested be held inviolate, or the disclosure of which would be embarrassing or would be likely to be detrimental to the employer or client. Information ceases to be confidential when the employer or client so indicates or when such information becomes publicly known.
- An archaeologist shall not:
  a. Reveal confidential information, unless required by law;
  b. Use confidential information to the disadvantage of the client or employer;

c. Use confidential information for the advantage of herself/himself or a third person, unless the client consents after full disclosure;

d. Accept compensation or anything of value for recommending the employment of another archaeologist or other person, unless such compensation or thing of value is fully disclosed to the potential employer or client;

e. Recommend or participate in any research which does not comply with the requirements of the Standards of Research Performance.

Website: *www.rpanet.org*

THE EUROPEAN ASSOCIATION OF ARCHAEOLOGISTS (EAA)

Principles of Conduct for Archaeologists in Contract Archaeological Work

The membership of the EAA voted to approve and adopt a set of Principles of Conduct for archaeologists involved in contract archaeological work. These had been prepared by the EAA's Working Party on Commercial Archaeology, were aired at the Ravenna meeting in 1997, and were published in draft in *The European Archaeologist* 8 (Winter 1997). The draft principles were further discussed at a well-attended and lively roundtable held at the Göteborg meeting.

The text that was approved by the membership is reproduced below. The Principles of Conduct help to define the standards of conduct expected of professional archaeologists in Europe.

Two important changes were made as a result of the discussions at Göteborg. First, the earlier phrase "commercial archaeological work" was replaced with "contract archaeological work." This reflects the view that archaeology is not, in the end, a commercial activity (even though it is often carried out under contracts, of various kinds). Secondly, a new principle (No. 14) was added. This reflects the importance of promoting both the principles and the means to make them work in practice. The need for adequate regulation of contract archaeology (normally by state or municipal authorities, but with professional associations also having a crucial role to play) is especially important.

Note: Many of these principles apply equally to all kinds of archaeological work, but this code deals especially with issues arising from a contract system of funding.

1. Archaeologists should ensure that they understand, and operate within, the legal framework within which the regulation of archaeological work takes place in that country.
2. Archaeologists should ensure that they give the best possible advice to developers and planners, and should not advise on matters beyond their knowledge or competence.
3. Archaeologists should ensure that they understand the structure of archaeological roles and responsibilities, the relationships between these roles, and their place in this structure.
4. Archaeologists should avoid conflicts of interest between the role of giving advice in a regulatory capacity and undertaking (or offering to undertake) work in a contract capacity.

5. Archaeologists should not offer to undertake contract work for which they or their organizations are not suitably equipped, staffed, or experienced.
6. Archaeologists should maintain adequate project control systems (academic, financial, quality, time) in relation to the work which they are undertaking.
7. Archaeologists should adhere to recognized professional standards for archaeological work.
8. Archaeologists should adhere both to the relevant law and to ethical standards in the area of competition between archaeological organizations.
9. Archaeologists involved in contract archaeological work should ensure that the results of such work are properly completed and made publicly available.
10. Archaeologists involved in contract archaeological work should ensure that archaeological information is not suppressed unreasonably or indefinitely (by developers or by archaeological organizations) for commercial reasons.
11. Archaeologists involved in contract archaeological work should be conscious of the need to maintain the academic coherence of archaeology, in the face of a tendency toward fragmentation under a contract system of organization.
12. Archaeologists involved in managing contract archaeological work should be conscious of their responsibilities towards the pay, conditions of employment and training, and career development opportunities of archaeologists, in relation to the effects of competition between archaeological organizations on these aspects of life.
13. Archaeologists involved in contract archaeological work should recognize the need to demonstrate, to developers and to the public at large, the benefits of support for archaeological work.
14. Where contract archaeology exists, all archaeologists (especially in positions of influence) should promote the application of this code, and promote development of the means to make it work effectively, especially adequate systems of regulation.

Website: *www.e-a-a.org/EAA_Princ_of_Conduct.pdf*

Appendix D
A Global Guide to Heritage Management

In this appendix we provide a country-by-country description of the way countries manage their historic and cultural monuments, places, archaeological remains, and architecture. Heritage management systems vary around the around. Some countries have national centralized programs, while many smaller countries, especially in the non-Western world, do not. In many countries, heritage professionals are government employees; in other countries, heritage management has been privatized and consists of private-sector heritage professionals with some governmental oversight or coordination.

For many of the smaller and less industrialized countries, finding information about heritage protection requirements can be difficult, especially if one is not familiar with the local language(s). As indicated in the country-by-country list below, government agencies with heritage resource responsibilities may use any number of words in their titles, including "archaeological," "monuments," "heritage," "cultural," and "history." In addition, heritage resource responsibilities may be embedded within larger organization—for example, those with environmental protection responsibilities. Further, depending on the cultural and historical significance of the project location and associated resources, governmental involvement may come at the local level, regional level, or national level.

Where project personnel are having trouble locating information about a country's protocols, one might check with national museums and universities, which can usually be found on the Internet. The About.Com: Archaeology website (*www.archaeology.about.com*). Cultural, historical, and ecological parks and preserves also may have staff that can provide information, contacts, and suggestions. Sites included in the ICOMOS World Heritage List are found in 145 different countries and should have knowledgeable staff available to consult. If a country has a World Heritage Site, we have indicated that below; each World Heritage site has an information sheet with contact information, which can be found at the World Heritage website (*http://whc.unesco.org/en/*). In addition, ICOMOS has reports from 54 countries concerning heritage resource at risk, which may provide additional insight useful for developers working in those countries (*www.international.icomos.org/risk/2002/nat_reports.htm*).

To organize the information, we have divided the world into five different regions:

- Western Hemisphere;
- Europe;
- Near East and South Central Asia;
- East Asia and Pacific;
- Africa.

The information provided below is largely from English-based sources, which has limited the information for countries included in this edition. We encourage those with experience in countries where we have provided limited or incorrect information to send corrections to us at our website: *NorthwestAnthropology.com*. As new information is received, we'll update this appendix and provide it on the website.

Where we have been unable to find specific information on requirements that projects must fulfill before beginning construction, we have indicated this with Unk(nown) in column 3. If we have obtained specific and useful information, rather than put this information in the body of the table, these will follow as text at the end of each regional table.

WESTERN HEMISPHERE

COUNTRY	WORLD HERITAGE SITES	REQUIRE-MENTS	ADDITIONAL WEBSITES
Argentina	8	Unk	
Belize	1	See text	www.ambergraiscaye.com/pages/town/parkintro.html
Bolivia	6	Unk	
Brazil	17	See text	www.iphan.gov.br
Canada	15	See text	www.pc.gc.ca
Chile	5	Unk	
Colombia	6	See text	www.icanh.gov.co/secciones/legislacion/decreto_icanh.htm (in Spanish)
Costa Rica	3	See text	www.museocostarica.go.cr/
Cuba	9	Unk	
Dominica	1	Unk	

COUNTRY	WORLD HERITAGE SITES	REQUIRE-MENTS	ADDITIONAL WEBSITES
Dominican Republic	1	Unk	
Ecuador	4	Unk	
El Salvador	1	Unk	
Guatemala	1	Unk	
Haiti	1	Unk	
Honduras	2	See text	
Mexico	29	See text	www.eccultura.gob.mx
Nicaragua	1	Unk	
Panama	5	Unk	
Paraguay	1	Unk	
Peru	10	See text	www.inc.gob.pe
U.S.	20	See text	See text
Uruguay	1	Unk	
Venezuela	3	See text	

Belize

Although no information was found concerning requirements that projects must fulfill before beginning construction, Belize has a rich archaeological record, and strong interest in environmental and ecological preservation so care should be taken. *www.ambergriscaye.com/pages/town/parkintro.html*).

Brazil

The Brazilian government has several laws to protect the heritage resources, as well as procedures for interventions (construction or otherwise) that may negatively impact archaeological and historical architectural sites. *www.iphan.gov.br*.

Canada

Canada established Parks Canada to develop rules and procedures for heritage resources on federal crown property. *www.pc.gc.ca*.

In addition, each Canadian province has its own specific rules and procedures overseen by its own heritage branch. Check the individual provincial websites for information concerning project requirements:

Alberta: *www.culture.alberta.ca/heritage/default*
British Columbia: *www.tca.gov.bc.ca/heritage*
Manitoba: *www.gov.mb.ca/chc/*
New Brunswick: *www.gnb.ca/013/heritage-e.asp*
Newfoundland and Labrador: *www.heritage.nf.ca/home.html*
Nova Scotia: *www.gov.ns.ca/tch/default.asp*
Ontario: *www.culture.gov.on.ca/english/heritage/index.html*
Prince Edward Island: *www.gov.pe.ca/cca*
Quebec (no website identified)
Saskatchewan: *www.tpcs.gov.sk.ca/heritage*

Colombia

The national authority on archaeological heritage is the Institute of Colombia Anthropology and History (ICANH). In order for a project to get an environmental permit from the Ministry of Environment (or from other lower-ranked offices, depending on the scale and projected impact of the construction), management will need to develop a Preventive Archaeology Program with the assistance of a professional archaeologist. Archaeological studies (excavations, surveys, etc.) required for the project need specific permits, which are issued by the ICANH. *www. icanh.gov.co/secciones/legislacion/decreto_icanh.htm* (in Spanish).

Costa Rica

Cultural heritage site protection in Costa Rica falls is authorized by the Defense and Conservation of Heritage Act (No. 6703). Protection is the responsibility of the Department of Archaeology and History within the National Museum of Costa Rica *www.museocostarica. go.cr/*.

Cuba

Cuba appears to have laws and procedures for protecting archaeological sites. However, no specific information has been found to date.

Honduras

The cultural heritage of Honduras is protected under the 1984 Law for the Protection of the Honduras Cultural Heritage, revised in 1997 (Decree 220-97). No information about procedures or who to contact, however, was found.

Mexico

The National Institute of Anthropology and History (INAH) enforces heritage and archaeological laws for Mexico. Refer to the website for review procedures and contact information (*www.eccultura.gob.mx*).

Peru

The National Institute of Culture (NIC), a government agency within the Ministry of Education, enforces heritage and archaeological laws for Peru. Projects will typically need an approval from a recognized archaeologist, which is then validated by the INC. *www.inc.gob.pe/*.

United States

Heritage management in the United States is coordinated by a variety of federal and state programs. The Advisory Council on Historic Preservation (ACHP) is an independent federal agency that promotes the preservation, enhancement, and productive use of our nation's historic resources, and advises the President and Congress on national historic preservation policy (*www.ACHP.gov*). The National Park Service coordinates a variety of heritage management activities, depending on whether the topic relates to collections, landscapes, archaeology, architecture, engineered structures, or interpretation (*www.nps.gov*). A list of significant buildings, archaeological sites, traditional use areas, engineered structures, and landscapes, referred to as the National Register of Historic Places, is maintained by the National Park Service.

Projects that involve federal funding, lands, or approvals require a heritage review to understand the potential effects that the project might have on important resources. Reviews are mandated under the National Historic Preservation Act Section 106, and require project proponents to resolve any adverse effects to sites listed on the National Register of Historic Places. The Section 106 review process is administered at the state level by the State Historic Preservation Officer (SHPO). Every State and U.S. territory has a SHPO who, with the support of qualified staff, are charged with providing consultation for federal undertakings under Section 106. To locate one of the 53 SHPOs and obtain information on their review procedures, check the website: *www.ncshpo.org/find/index.htm*.

Seventy-six Tribal Historic Preservation Officers (THPO) administer heritage resource compliance activities on 34 million acres in 24 states. To locate one of the THPOs and obtain information on their review procedures, check the website: *www.nathpo.org/map.html*.

Project reviews are also required under the National Environmental Policy Act (NEPA). These assessments are used in environmental assessments and environmental impact statements to assist in the evaluation of project alternatives. Where impacts from the preferred alternative are unavoidable, mitigation actions are developed. Additional information on complying with NEPA requirements is found on the Council of Environmental Quality website: *www.whitehouse.gov/administration/eop/ceq/*.

Venezuela

Protection of archaeological sites in Venezuela is mandated by the Law of Protection and Defense of the Cultural Heritage (Gaceta Oficial #4.623: 1993). No additional information about governmental oversight has been found.

EUROPE

Heritage management is a mature industry throughout most of Europe. Information about heritage management in parts of Europe was found in Discovering the Archaeologists of Europe, a transnational project, which produced twelve country-specific reports on the archaeological profession in that country (*http://discovering-archaeologists.eu/*). The countries with such reports are indicated in the table below.

Developers have had to accept greater responsibility for archaeological protection since the early 1990s, when The Convention on the Protection of the Archaeological Heritage of Europe, usually referred to as the Valletta Treaty or Malta Convention, was signed by some European countries.

Other international agreements on the protection of cultural heritage signed by various European countries include the following:

- the Hague Convention for the Protection of Cultural Property in the Event of Armed Conflict
- the Convention Concerning the Protection of World Cultural and Natural Heritage (UNESCO Convention)
- the Convention for the Protection of the Architectural Heritage of Europe (Granada Convention)
- the European Convention for the Protection of the Archaeological Heritage (the La Valletta Convention)
- the European Landscape Convention.

Details on heritage legislation across Europe is found on the website: *http://heritagelaw.org/*. Any conventions signed by specific countries are noted in the table.

EUROPE

COUNTRY	WORLD HERITAGE SITES	REQUIRE-MENTS	DAE REPORT*	TREATIES	ADDITIONAL WEBSITES
Austria	5	See text	Yes		
Azerbaijan	5	Unk			
Belarus	4	Unk			
Belgium	9	See text	Yes	M/V 2002†	
Bosnia/ Herzegovina	2	Unk			
Bulgaria	9	Unk			

Country	World Heritage Sites	Require-ments	DAE Report*	Treaties	Additional Websites
Croatia	7	Unk			
Cyprus	5	See text	Yes		
Czech Rep.	12	See text	Yes		
Denmark	4	See text			
Estonia	2	Unk			
Finland	7	Seetext			www.nba.fi/en/archaeology
France	33	See text			
Georgia	3	Unk			
Germany	33	See text	Yes	M/V 2002†	
Greece	17	See text	Yes		
Hungary	8	See text	Yes		
Iceland	2	See text			http://eng.umhverfisraduneyti.is
Ireland	2	See text	Yes		www.archaeology.ie.en
Italy	43	See text			www.beniculturali.it/mibac/export/MiBAC/index.html
Latvia	2	Unk			
Lithuania	4	Unk			
Luxem-bourg	1	Unk			
Malta	3	Unk			
Moldava	1	Unk			
Montenegro	2	Unk			
Netherlands	7	See text	Yes	M/V†	

COUNTRY	WORLD HERITAGE SITES	REQUIRE-MENTS	DAE REPORT*	TREATIES	ADDITIONAL WEBSITES
Norway	7	See text			www.environment.no/tema/kulturminner/
Poland	13	See text			
Portugal	13	See text			
Romania	7	Unk			
Russian Fed	23	Unk			
San Marino	1	Unk			
Serbia	4	Unk			
Slovakia	7	See text	Yes		
Slovenia	1	See text	Yes		
Spain	40	See text			
Sweden	14	See text			See text
Switzerland	9	See text			
Ukraine	4	Unk			
United Kingdom	27	See text	Yes		www.direct.gov.uk/en/index/htm

*DAE Report = Discovering the Archaeologists of Europe (see *http://discovering-archaeologists.eu/*
†M/V = Malta Convention / Valletta Treaty

Austria

In Austria, protection of cultural heritage (*Denkmalschutz*) is a responsibility of the central state. Although we found no specific laws and procedures for projects to follow, we did learn that contract archaeology is an emerging field in Austria, in response to an increasing influx of private money from building companies to finance large-scale excavations; allowing developers to fund the recovery of archaeological remains within the project area enables the developer to control the schedule, as opposed to prior times when developers had to wait for the government archaeological services to complete the archaeological work. Austria has not yet ratified the Valletta convention (European Convention for the Protection of the Archaeological Heritage) and thus has not introduced polluter financing into archaeology as a legal requirement.

Belgium

Responsibilities for heritage management are conducted at the regional and community level. As a result there are a large number of institutions which may have a responsibility for protecting Belgium's heritage, depending on the location and the nature of the resource involved. Private-sector archaeology is in its infancy in Belgium. Belgium signed the European Convention on the Protection of the Archaeological Heritage (the Malta/Valletta Convention) in 2002 and expects developers to ensure protection of the cultural environment from project activities.

Cyprus

The archaeological heritage of Cyprus falls under the authority of the Department of Antiquities, and largely involves facilitating large archaeological excavations.

Czech Republic

Archaeological and heritage resources in the Czech Republic are managed at the regional level; there are fourteen regions. Private archaeological companies began in the early 1990s and are concentrated in the major building areas (Prague, Brno, and Plzen).

Denmark

The Heritage Agency of Denmark is responsible for preserving the historic buildings, sites and monuments, and museum collections that are part of Denmark's cultural heritage. The Heritage Agency regularly participates in national and international projects for the mapping, preservation, and dissemination of Denmark's cultural heritage. The Heritage Agency is a body under the Danish Ministry of Culture.

Finland

The National Board of Antiquities preserves Finland's material cultural heritage: collects, studies, and distributes knowledge of it. The Board is a cultural and research institution, but it is also a government authority charged with the protection of archaeological sites, built heritage, cultural and historically valuable environments and cultural property, in collaboration with other officials and museums. The National Board of Antiquities of Finland is attached to the Ministry of Education. For more information, visit *www.nba.fi/en/archaeology*.

France

Archaeological matters fall under the purview of the French Ministry of Culture. Private-sector archaeology does not exist in France.

Germany

Heritage management in Germany is broad, complex, and decentralized. There are private firms that assist with archaeological excavations and provide other archaeological services.

However, in most areas, the state coordinates and conducts any archaeological work that needs to be done.

Greece

Greece has had a long tradition of protecting its archaeological and other heritage resources. Archaeologists and other heritage professionals are found throughout the country, in a variety of state institutions. Commercial archaeology does not exist in Greece. Rather, all archaeological activities, and especially excavations, are undertaken by the governmental Archaeological Service organization. Sometimes, other scientific institutions (such as Departments of Universities or Foreign Archaeological Schools) obtain special permits from the Archaeological Service organization and conduct their own excavations.

Hungary

Development project personnel will likely deal with an archaeologist or other heritage specialist from the Office for Cultural Heritage, which has had a country-wide supervisory role over archaeological sites in Hungary. Depending on the situation and resources involved, heritage professionals from the Field Service for Cultural Heritage or the Ministry of Education and Culture may be involved. There are a small number of private-sector archaeologists, who primarily participate in large-scale excavations.

Iceland

Affairs associated with cultural heritage in Iceland fall under the purview of the Ministry for the Environment. For information concerning any work being conducted involving the environment of Iceland, the Ministry of Environment website contains information regarding legislation, information, and agencies: *http://eng.umhverfisraduneyti.is*.

Ireland

The National Monument Service oversees heritage matters in Ireland. The website contains the various rules, regulations, and forms involved in protecting and managing the Irish heritage (*www.archaeology.ie/en/*). Project reviews involve consulting Record of Monument and Places (RMP) maps to determine whether the proposed project lies in proximity to a recorded monument. These maps mark the locations of over 120,000 recorded monuments. The RMP is available in planning authority offices and in public libraries countrywide as well as in county and city/town museums. It also may be consulted at the National Monuments Archive, 6 Ely Place Upper, Dublin 2.

Italy

Cultural heritage in Italy is active and diverse. Procedures and contacts vary depending on location and type of resources. The Ministry for Cultural Heritage and Activities is a good starting point: *www.beniculturali.it/mibac/export/ MiBAC/index.html*.

Netherlands

Almost all government departments in the Netherlands have some level of responsibility for heritage management. However, recent legislative developments have delegated many responsibilities to provinces, local governments, and the private sector, which has begun to play an increasingly important role. Private-sector involvement in archaeology is a relatively new phenomenon and is a direct consequence of the adoption of the Malta/Valletta Convention.

Many archaeological matters fall under the purview of the Monuments Act. Projects initiated by the private sector are expected to identify any heritage resources in the project area and ensure protection if there will be impacts. Government bodies make archaeological protection a condition for development permits. If archaeological mitigation is needed, the developer is responsible for the costs.

Norway

Norway has a strong cultural heritage management program. The Directorate for Cultural Heritage in Norway (Riksantikvaren) is responsible for the practical implementation of the Norwegian Cultural Heritage Act and the objectives laid down by the Norwegian Parliament (Stortinget) and the Ministry of the Environment. Proposed developments are reviewed by heritage staff in order to minimize destruction of the cultural environment. For information concerning archaeological monuments, archaeological heritage, cultural environments, world heritage, international agreements, and cultural heritage in Norway, consult the Cultural Heritage website: *www.environment.no/tema/kulturminner/*.

Poland

Cultural heritage protection in Poland is authorized by the Protection and Guardianship of Monuments Act (23 July 2003). Authority for protection is distributed among the following organizations in the Ministry of Culture: The Department for the Protection of Monuments and the Department of National Heritage.

Portugal

Cultural heritage protection is authorized under the Portuguese Heritage Protection Law (Law 107/2001).

Slovakia

The Slovak Republic is divided into eight regions with state authorities responsible for cultural heritage protection. Cultural heritage protection responsibilities are assigned to the Ministry of Culture of the Slovak Republic, the Monuments' Board of the Slovak Republic, and regional Monuments' Boards. Developers pay most costs associated with protecting heritage resources affected by their projects. A major motorway construction project in the 1990s was instrumental in formulating the way Slovenia will protect resources from development. Private companies have formed in recent years to assist commercial developments that have encountered archaeological deposits.

Slovenia

Heritage matters in Slovenia are addressed within the Institute for the Protection of Cultural Heritage.

Spain

Heritage sites in Spain are under the purview of the Institute of Cultural Heritage of Spain, an organization within the Directorate General of Fine Arts and Cultural Assets of the Ministry of Culture.

Sweden

Heritage and historic environment issues are the responsibility of the Swedish National Heritage Board, under the auspices of the Ministry of Culture. The objectives for the Board include encouraging: (1) preservation and protection of the historic environment; (2) respect for the heritages of different groups; (3) appreciation of, commitment to, and the assumption of responsibility for one's own heritage. *www.raa.se/cms/extern/en/about_us/our_mission/our_ mission.html.*

Switzerland

Cultural heritage protection in Switzerland is mandated by the Law and Decree about Nature and Heritage Preservation (Article 78 of the Swiss Federal Constitution). Two commissions responsible for protecting the county's heritage are the Federal Commission for Nature and Heritage Preservation and the Federal Commission for Historic Monuments and Site Preservation.

United Kingdom

In the United Kingdom, developers are expected to fund archaeological evaluation and mitigation as part of the process required to support for their construction project. The developer will need to apply to the local government for planning approval; the local government will have archaeological expertise to evaluate the findings and plans of the developer. Private sector heritage consultants and commercial archaeological contractors are available in the United Kingdom to provide advice to developers and conduct the work.

For all projects being conducted in the United Kingdom, please check the Directgov website for all information concerning rules, regulations, forms, and general information: *www.direct.gov.uk/en/index.htm*.

NEAR EAST AND SOUTH CENTRAL ASIA

Little heritage management information about the Near East and South Central Asia has been collected, primarily due to language barriers. Other than the number of World Heritage sites in each country, we were only able to obtain information on Egypt and India.

World Heritage sites in this region include the following: Afghanistan (2), Albania (2), Andorra (1), Armenia (3), Bahrain (1), Bangladesh (3), Egypt (7), India (23), Islamic Republic of Iran (9), Iraq (3), Israel (6), Jordan (3), Kazakhstan (3), Lebanon (5), Libyan Arab Jamahiriya (5), Nepal (4), Oman (3), Pakistan (6), Saudi Arabia (1), Sri Lanka (7), Sudan (1), Syrian Arab Republic (5), Tunisia (8), Turkey (9), Turkmenistan (3), Uzbekistan (4), and Yemen (4). Please check *http://whc.unesco.org/en* for further information.

Egypt

Egypt currently has no laws requiring project backers to make sure no impacts to important resources are caused by project activities. The Egyptian Cultural Heritage Organisation (ECHO) formed in 1996 to facilitate protection of Egypt's cultural heritage. Among its goals is developing a salvage program to combat the threat of building and industrial development: *www.e-c-h-o.org/index2.htm*.

India

Cultural heritage in India is under the purview of the Archaeological Survey of India, under the Ministry of Culture (*http://asi.nic.in/asi_aboutus.asp*). While India's program is diverse and extensive, no specific rules to ensure protection of important resources from development and construction could be found.

AFRICA

Little heritage management information about central Africa has been collected, primarily due to language barriers. Some information on southern Africa was obtained from the Association of Southern African Professional Archaeologists, which presents information on its website, *www.asapa.org.za/*. What little information we were able to find is presented below.

World Heritage sites in this region include the following: Algeria (7), Benin (1), Botswana (1), Cameroon (1), Central African Republic (1), Côte d'Ivoire (3), Democratic Republic of the Congo (5), Ethiopia (8), Gabon (1), Gambia (2), Ghana (2), Guinea (1), Kenya (4), Madagascar (3), Malawi (2), Mali (4), Mauritania (2), Mauritius (2), Morocco (8), Mozambique (1), Namibia (1), Niger (1), Nigeria (2), Senegal (5), Seychelles (2), South Africa (8), Tanzania (1), Togo (1), Uganda (3), Zambia (1), Zimbabwe (5). Check the website for additional information: *http://whc.unesco.org/en*.

Lesotho

Heritage resource management in Lesotho is authorized under the Historical Monuments, Relics, Fauna and Flora Act, Act 41 of 1967 (*www.asapa.org.za/images/uploads/Lesotho.pdf*). Developers should contact the Department of Culture, Ministry of Tourism, Environment and Culture (*www.mtec.gov.ls/culture/*) to determine if any national or provincial sites, heritage areas or protected areas will be affected by project activities. The Association of Southern African Professional Archaeologists website presents information on its website that may be useful to project managers working in Lesotho: *http://www.asapa. org.za/*.

Mozambique

Heritage resource management in Mozambique is administered by the National Directorate for Culture of Mozambique, Ministry of Education and Culture (Rua Capitão Henrique de Sousa N. 15, P.O.Box 1742, Maputo, Mozambique).

The Association of Southern African Professional Archaeologists website presents information on its website that may be useful to project managers working in Mozambique (*www.asapa.org.za/*). There is one World Heritage site in Mozambique; check the website for additional information: *http://whc.unesco.org/en*.

Namibia

Heritage resource management in Namibia is authorized under the National Heritage Act, Act 27 of 2004 (*http://rockart.wits.ac.za/misc/Heritage_Act_2004. pdf*). A national campaign, Heritage Hunt, is underway to identify and protect Namibia's cultural heritage.

Developers should contact the Heritage Sites Project Coordinator, to determine if any national or provincial sites, heritage areas, or protected areas will be affected by project activities (Museums Association of Namibia, PO Box 147, Windhoek [tel/fax: +264(0)61-302236; cell: +264(0)812805130; email: *museums@iway.na*]).

The Association of Southern African Professional Archaeologists website presents information on its website that may be useful to project managers working in Namibia: *www.asapa.org.za/*.

South Africa

Heritage resource management in South Africa is authorized under the National Heritage Resources Act of 1999 (*www.acts.co.za/ntl_heritage_res/ index.htm*). Developers should contact the South Africa Historic Resources Agency in Cape Town (*www.sahra.org.za/default.asp*) to determine if any national or provincial sites, heritage areas or protected areas will be affected by project activities. The Association of Southern African Professional Archaeologists website presents information on its website about South African resources, laws, and contacts: *www.asapa.org.za/*.

Zambia

Heritage resource management in Zambia is authorized under the National Heritage Conservation Commission Act of 1994 (*www.asapa.org.za/images/uploads/Zambia.pdf*). Developers should contact the National Heritage Conservation Commission in Livingston or one of the four regional offices (P.O. Box 60124, Livingston, Zambia) to determine if any national or provincial sites, heritage areas or protected areas will be affected by project activities. The Association of Southern African Professional Archaeologists website presents information on its website that may be useful to project managers working in Zambia: *www.asapa.org.za/*.

Zimbabwe

Heritage resource management in Zimbabwe is authorized under the National Museums and Monuments Act No. 17 of 1972. Developers should contact the National Museums and Monuments of Zimbabwe, Ministry of Home Affairs in Harare to determine if any national or provincial sites, heritage areas or protected areas will be affected by project activities: *www.zimheritage.co.zw/site/index.html*. The Association of Southern African Professional Archaeologists website presents information on its website that may be useful to project managers working in Zambia: *www.asapa.org.za/*.

East Asia and Pacific

Little heritage management information about the East Asia and the Pacific has been collected. World Heritage sites in this region include the following: Australia (17), Cambodia (2), China (37), Indonesia (7), Japan (14), Democratic People's Republic of Korea (1), Lao People's Democratic Republic (2), Malaysia (3), Mongolia (2), New Zealand (3), Papua New Guinea (1), Philippines (6), Republic of Korea (8), Solomon Islands (1), Thailand (5), Vanuatu (1), and Vietnam (5).

Australia

Australia is among the world leaders in heritage management, especially in the area of indigenous people's rights, as a result of the Aboriginal Cultural Heritage Act of 2003 (see *www.environment.gov.au/heritage/*). Management is coordinated at the state level, as explained at the respective websites:

Queensland: *www.environment.gov.au/heritage/places/qld/index.html*

New South Wales: *www.environment.gov.au/heritage/organisations/ state/nsw.html*

Northern Territory: *www.environment.gov.au/heritage/places/nt/index.html*

South Australia: *www.environment.gov.au/heritage/places/sa/index. html*

Victoria: *www.environment.gov.au/heritage/places/vic/index.html*

Western Australia: *www.environment.gov.au/heritage/places/wa/ index.html*

Tasmania: *www.environment.gov.au/heritage/places/tas/index.html*

Australian Capital Territory: *www.environment.gov.au/heritage/places/act/index.html*.

Australia has a strong private-sector archaeology. Refer to the Australian Association of Consulting Archaeologists website for information or to locate a professional archaeologist or heritage consultant: *www.aacai.com.au/*.

China

Little information about heritage management in China was found. A newsletter about heritage management in China is produced by The Australian National University (*http://www.chinaheritagenewsletter.org/index.php*). A recent book, *Cultural Heritage Management in China: Preserving the Cities of the Pearl River Delta*, edited by Hilary du Cros, Yok-shiu F Lee (2007 Routledge), focuses on urban sites.

Japan

Although we found little information concerning Japanese heritage management, a substantial summary of archaeology in Japan is found at *www.t-net.ne.jp/~keally/jpnarch.html*. In general, all construction has to be reported to the local authorities before even serious planning can go forward. That authority often will be the local city hall. They will direct the company to the cultural properties section (e.g., bunkazai tanto). They will go over their maps of known site locations and let the contractor know if there is a site already known on the planned construction site. They will also judge the possibility of a site being there that has not yet been found. Then, except for places lacking any site or other cultural property, they will start with a test excavation, paid for by the contractor. The results of that test will determine what happens next—construction goes on or a full excavation is conducted before construction, or something in between, all paid for by the contractor. Full excavation in Japan entails 100% of the site area that will be disturbed by the construction.

New Zealand

The cultural heritage of New Zealand is under the purview of the Department of Conservation. No specific rules for developers were found for New Zealand, however, the Department does have an active program managing known resources: *www.doc.govt. nz/conservation/historic/*.

Philippines

Preserving Filipino cultural heritage is the responsibility of the National Commission for Culture and Art (NCCA) and the National Historical Institute (NHI). However, there are no laws or regulations requiring protection of heritage sites.

Glossary:
Archaeological, Heritage Resource Management, and Project Management Terms

Absolute Dating – Determining the age of an object in calendar years, as opposed to *relative dating*. A date for an archaeological site can be determined using scientific methods such as tree-ring dating, radiocarbon dating, and thermoluminescence.

Activity Area – The portion of a site in which a specific activity took place.

Adverse Effect – When a project activity causes substantial disturbance to an archaeological site or other heritage resources and destroys the very qualities of the resource that make the resource important, the disturbance is said to have had an adverse effect. A "Determination of Adverse Effect" is one of the steps in the National Historic Preservation Act Section 106 process.

Advisory Council on Historic Preservation (ACHP) – A U.S. government organization established to advise the President on historic preservation issues. The ACHP is an advisory council and plays a significant role in the National Historic Preservation Act Section 106 process.

Aerial photography – Photographs taken from above in an aircraft. This can be useful for identifying any number of aspects about a project area and archaeological resources located there. Vegetation changes, landforms, foundations, and roads are often observable.

Anthropology – The study of the origins and diversity of the human species. In the United States, the discipline is composed of four subfields: biological anthropology, social anthropology, linguistics, and archaeology.

Arbitrary Level – Archaeologists will often excavate in fixed blocks (e.g., 5–10 cm at a time) in order to maintain vertical control of where artifacts are found within an archaeological excavation unit.

Antiquities – Objects of sufficient age to be valued for their age and character.

Antiquities Act of 1906 – The first U.S. law to attempt protection of archaeological sites. This act gave the President authority to establish national monuments based on archaeological resources and also made looting illegal on federal lands.

Archaeological Records – Archaeological records generally refer to the various metadata generated before, during, and after an archaeological investigation. Records can include field notes, photographs, maps, drawings, tabulations, analyses, computer files, interviews, budgets, expenditures, and all administrative materials. It is critical that the archaeological records be preserved with the archaeological collections so that future researchers will be able to reconstruct the excavated materials from the site and extract information.

Archaeologist – An individual trained in the discipline of archaeology. Some countries have established prerequisites that a person must have in order to be considered a professional archaeologist (e.g., college degree, years of experience). Some may be labeled field archaeologists, who are highly skilled in excavation methods; others may have leadership skills and serve as crew chiefs or principal investigators; still others may be skilled in identifying and analyzing particular artifact types such as stone tools, ceramics, or faunal remains.

Archaeology – Archaeology is the study of past human cultures through the discovery and interpretation of materials remains (artifacts).

Archaeological Assessment – In the context of this book, an archaeological assessment is a statement concerning the potential for a project to encounter and impact archaeological remains. An assessment can range from a brief, initial cursory look that draws on an archaeologist's knowledge of the area, to a detailed, thoroughly researched and well-sampled investigation based on hard facts.

Archaeological Site Characterization – Efforts undertaken to describe the horizontal and vertical extent of an archaeological site and understand the nature of the deposits.

Archival Research – A search for information, usually documentary, that relates to the project area under consideration. Old maps, land records, histories, photographs, government records, biographies, etc., found in libraries, government agencies, and archives can provide invaluable assistance to the archaeologist trying to understand what might be located in the project area. Also sometimes referred to as *background research* or *a file and literature search.*

Area of Potential Effect (APE) – A term used in National Historic Preservation Act Section 106 compliance meaning the area that could be impacted by a construction project, and therefore the area that will be studied for heritage resources. Defining the APE is the initial step in the Section 106 process.

ARPA – Archaeological Resources Protection Act of 1979 – A U. S. federal law that protects archaeological sites that are older than 100 years on federal and Indian land. Looting and theft are felony crimes, and a first offense can bring a fine and one year in prison. This protection does not extend to sites on private land.

Arrowhead – A projectile point that was hafted to an arrow shaft and used with a bow.

Artifact – An object that has been made or used by people. Artifacts, for example, can be stone tools (the finished product), the *flakes* (by-products) that were generated during the manufacture of the stone tool, and the butchered bones that resulted from using the stone tool.

Auger — A tool used to dig subsurface samples in order to establish the presence or absence of archaeological materials. Using an auger is a quick way to check subsurface deposits across a large area; however, they are small and may miss deposits. If the soils are rocky, augering is not effective.

Background Research – Typically conducted prior to archaeological fieldwork to determine what other work has been done in and around the location, what is known about the history of the area, and what is known about the environment. See *Archival Research*.

Biface – A lithic artifact with flake scars on both sides.

Bioturbation – The mixing of soils resulting from plant and animal activity.

Burial – Intentionally buried human remains (see Sprague 2005).

Burial Ground – An area set aside typically for human interments, such as a cemetery.

C-14 Dating – Also known as carbon-14, a radiometric technique developed in the 1950s that enables archaeologists to get an absolute calendrical age from charcoal, bone, shell, or wood.

Cache – Protected storage area for culturally important materials such as food, tools, and medicines for future use.

Cemetery – A location set aside by a family or community for burying the deceased (see Sprague 2005).

Characterization – An analysis of a project area or archaeological site by sampling the surface and subsurface in order to describe the nature and location of various attributes.

Chronology – A sequential ordering that places cultural entities in time.

Community Development – Project in a community proposed or funded to promote economic growth and increased quality of life.

Conceptual Model – Based on the *characterization* of a site or project area, the hypothesized locations of archaeological deposits, the associated geomorphology, and the areas disturbed by subsequent disturbance. Conceptual models are used to develop sampling strategies in order to test hypotheses and refine other conceptual models.

Consultation – Discussions held between one group and another to explain a proposed project and solicit issues, concerns, and expectations. Consultations may be government-to-government, meaning governmental officials are involved in the discussions.

Context – The space and time associations of an artifact or assemblage; for archaeologists, artifacts are rarely informative without data about the context in which they are found.

Contract Archaeology – Archaeological research conducted for pay, usually for some form of community development.

Core – A piece of tool stone from which pieces have been struck off during the process of tool manufacture.

Cortex – The original and outer surface of a stone.

Critical Path – A project management term referring to an activity or set of activities that must be completed before the next set of activities can begin.

Cultural Context – The cultural, historical, and environmental setting in which archaeological deposits or other heritage resources fit.

Cultural Landscape – An area that takes on contemporary meaning to a group of people for the resources that can be found there or events that occurred there in the past.

Cultural Resources – (also heritage resources) – A term used in the United States to refer to material and nonmaterial aspects of culture. It includes all time periods that are significant to living cultures, including groups maintaining and preserving their traditions. Originally developed to contrast with the concept of natural resources; today it is recognized that natural resources on which humans have relied (or continue to rely upon) are also cultural resources.

Cultural Resources Management (CRM) – A profession common in the United States relating to the preservation of cultural resources. It was developed to assist the government and the private sector to comply with legislation passed to protect the nation's heritage.

Culture History – The history of human settlement in a particular area or region. It focuses on placing cultures in a chronological framework.

Data Recovery – A term generally associated with excavation and analysis of archaeological materials commonly used in the context of archaeological mitigation.

Datum – A location designated at an archaeological site from which all measurements are taken.

Debitage – Lithic detritus resulting from tool manufacture. Useful to determine techniques and for showing technological traits. Debitage flakes usually represent the various stages of progress of the raw material from the original form to the finished product.

Diagnostic Artifact – An artifact that holds information about the time period, the cultural group, its function, or the activities that occurred onsite.

Direct effects – Impacts that occur immediately as a result of a project—for example, ground disturbance within an archaeological site.

Dispute Resolution – Procedures identified in agreement documents for the steps to follow when parties cannot agree on key actions.

Disturbance – Used in the context of an archaeological site that has been impacted by subsequent settlement, the concept is often used to argue that the archaeological deposit has been destroyed. More often than not, however, disturbance is limited to upper levels or certain portions of the project area, and intact archaeological deposits do exist.

Due Diligence – Conducting background checks and gathering current information about a piece of property or a potential employee that will reduce the risk of an archaeological disaster.

Ethnoarchaeology – The use of contemporary information from a group of people to aid in the interpretation of archaeological assemblages.

Ethnographic Landscape – An area that has special significance to a contemporary cultural group. The location may be associated with objects of importance, historical events, or important people.

Ethnohistoric Site – Refers to a place that has both cultural and historic ties to a particular group, such as a place where a particular group used to go to collect resources or conduct some activity.

Ethnohistory – The study of a particular cultural group of people by using information from oral histories and documentary references.

Excavation – Scientific removal of archaeological deposits using archaeological methods.

Faunal Remains – Animal bones found in archaeological contexts. Remains can refer to anything remaining of the animal after death.

Faunal Remains Analysis – A study of the animal remains (bones and shell) in an archaeological context. This can include an inventory, description, and quantitative assessment. The results provide behavioral information about past subsistence and the environment.

Feature – Something that is different from its surroundings. It may be a group of artifacts or simply a soil stain or remains of human activity that has an unknown function. Features are usually nonportable evidence of human behavior, such as fire hearths, house pits, garbage dumps, architectural elements, and surrounding matrix or soil stains.

Field Notes – All written documentation that is prepared at the project or archaeological site. These included handwritten notes, photographs and photo records, and illustrations (sketches, artifact drawings, or maps).

Fire-cracked Rock (FCR) – Rock that has been exposed to fire cracks in irregular patterns. Rocks that have been fractured by fire are angular and often reddened. FCR can suggest cooking and camping areas.

First Nations – Canadian term for the indigenous aboriginal communities.

Flakes (primary, secondary, tertiary) – By-products from flintknapping activities. A piece of stone removed from a larger mass by the application of force. Primary, secondary, and tertiary refer to the stages of manufacture when producing a flaked stone tool.

Flintknapping – The technology or art of stone tool manufacture.

Geofact – A naturally formed stone that looks like an artifact.

Geomorphology – The study of development and changes in landforms due to erosion and climate.

GPS (Global Positioning System) — A handheld device that pinpoints its own location-based satellite transmissions.

Ground-Penetrating Radar – A remote sensing technology that sends a radar pulse into the ground to identify unknown features.

Habitation Site – A location that has been or is lived in by humans.

Hammerstone – A stone used to strike flakes from a core.

Hearth – A fire pit used for heating or cooking.

Heritage Resources – An inclusive term to define all things cultural. Cultural and archaeological resources would fall under this term.

Historic Period – The period for which written records exist (varies throughout the world).

Historical Archaeology – The study of human behavior through material remains, using written history to aid in the interpretation.

Historical Architecture – Architecture that has taken on historic interest and may be old enough for protection under the law.

Historical Landform Analysis – Analysis performed by looking at old maps, aerial photographs, and historical photographs to identify changes that have occurred through time.

Historic Property – Any district, site, building, structure or object included in or eligible for inclusion in the National Register of Historic Places (16 U.S.C.470w[3]).

Historic Preservation – Preserving history by managing heritage resources. In the U.S., the National Historic Preservation Act was written to locate, document, manage, and protect these resources.

Historic Site – A cultural site that was occupied during a period in which written records exist. This period varies with geography.

House Pit – The remains of a prehistoric semi-subterranean house.

ICOMOS – International Council on Monuments and Sites, an international non-governmental organization of professionals, dedicated to the conservation of the world's historic monuments and sites (*www.icomos.org/*).

Impact – The effect that an activity has on a heritage resource.

Indirect Effects – The cumulative effects that an activity has on a heritage resource and will have for generations to come.

Infrastructure – Those parts of a construction project such as water lines, sewer lines, gas lines, electric lines, roads, and parking areas.

In Situ – A Latin term meaning "in place." Refers to an artifact that is documented within its original undisturbed location (e.g., the artifact was documented "in situ").

Inadvertent Discovery – The unearthing of archaeological deposits or human remains usually during construction. In the United States, the laws that protect human remains refer to inadvertent discoveries as buried human remains that were unknown until they were unintentionally discovered.

Indigenous Resources – Resources specific to those people who are native to the lands.

Industrial Archaeology – The study of industrial sites.

Interested Party – An individual or group that has ties to heritage resources within a project's area of potential effect or an individual or group that has an interest in the protection of those resources.

Inventory – A list of cultural resources that exist at an archaeological site.

Lay-down Area – An area designated by project management to store equipment or building materials before and during construction.

Level – An archaeological term representing layers of soil or cultural material.

Lithic Technology – The knowledge possessed by an individual or group allowing them to form stone into useful cutting, chopping, scraping, and pounding tools.

Lithic Artifacts – Tools and by-products made from stone material.

Looting – Unauthorized removal of cultural artifacts from any place. A looter is one who destroys archaeological sites in order to collect the artifacts for personal satisfaction or for sale.

Material Culture – The physical objects that represent a culture or a specific group of people.

Matrix – Archaeologically speaking, the sediments that surround and contain an artifact or site. For management purposes, a matrix is a tool to track risk management.

Memorandum of Agreement (MOA) – A pact made among different parties agreeing to certain actions; in the Untied States, MOAs are required under the National Historic Preservation Act if it has been determined that there will be an adverse effect from project activities on an important site.

Midden – An accumulation of organic matter (shell, bones) or other objects or debris discarded by inhabitants of a site.

Monitoring – Observing construction activity by archaeologists or cultural resources staff to identify and document cultural materials not visible prior to ground disturbance.

Mounds – Evidence of cultural activity. Usually mounding of soil for cultural purposes.

Native American Graves and Repatriation Act of 1990 (NAGPRA) – A U.S. law that established a process for returning human remains and items of cultural patrimony to indigenous peoples, and established procedures for handling inadvertent discoveries of human remains.

National Historic Preservation Act of 1966 (NHPA) – A U.S. law that, among other things, established a process for important heritage resources to be considered in planning which involves federal funding or federal land.

Native American – People indigenous to the Americas. Synonymous with American Indian.

Occupation – The use of a site or larger area by a cultural group.

Oral history – Spoken accounts of history, traditional cultural knowledge, experiences and stories that have been passed down through the ages. Oral histories may not be available to outside groups due to cultural sensitivity.

Palynology – The study of plant pollen for purposes of reconstructing past human environments.

Pedestrian Survey – A systematic walkover of a project area to identify and record cultural materials

Peer Review – A formal review of a publishable article or document by experts in that field of study.

Pestle – A food-grinding tool made of stone which is often longer than it is wide.

Petroglyph – A form of rock art that was pecked or incised into a stone.

Pictograph – Rock art that consists of pigment that is applied to a stone.

Pit house – A semi-subterranean house form with a superstructure made of perishable material such as reeds, hides, or wood.

Post hole – The previous location of a base of a pole or post, consisting of discolored soil and often associated with house and corral features.

Pothunting – See *Looting*.

Potsherd – A fragment of a ceramic pot or vessel.

Prehistoric Site – A location with archaeological materials that predates written records.

Predictive Model – A model that incorporates environmental and cultural information to identify areas that are likely to contain archaeological sites.

Process – A series of actions and interactions that brings about change.

Procurement – The act of going to get something (e.g., hunting, gathering, or getting money for a contract).

Project Area – The location of a project which includes the footprint of the construction project and the outlying areas that are associated with the project area in some way.

Projectile Point – A spear point, dart point, or arrowhead formed by chipping or flaking stone.

Provenience – The specific location of an artifact or feature documented by measurement in relation to a datum or a fixed point.

Quarry – A place where tool-grade stone occurs naturally.

Reconnaissance – One of the first stages of fieldwork, consisting of a systematic search for sites and artifacts via pedestrian survey.

Repatriate – Usually refers to human remains and their return to their original location.

Relative Dating – Dating by determining the sequence of components without the use of a calendar. (See *Absolute Dating.*)

Remote sensing – Using techniques such as aerial photography, magnetometry, and radar to help locate buried sites and features.

Risk – A concept that incorporates the likelihood of an event occurring and the consequences that will result should the event occur.

Risk Assessment – A formal process for determining the likelihood and consequences of various events important in the execution of a project.

Risk Management – A formal process for identifying actions project personnel can take to reduce the likelihood and consequences of a particular event.

Rock Art – See *petroglyph* and *pictograph*.

Sample – The studied portion of a site, feature, or cultural level.

Scraper – Usually refers to a formed tool that has been worked on a lateral edge or end and used to scrape organic material

Screening – An activity associated with excavation where shovelfuls of dirt are sifted through a quarter-inch or eighth-inch hardware mesh screen to recover small flakes, bones, and other remains of the site.

Section 106 – A requirement in the United States National Historic Preservation Act mandating that government agencies assess the effects of their actions on important cultural and historic resources.

Settlement Pattern – Also associated with subsistence; the overall land use patterns of a given cultural group, concerning traveling from place to place for hunting, gathering, trade, or other subsistence practices.

Shovel Testing – A method of excavation where small shovel holes are dug to a certain depth to determine presence or absence of subsurface archaeological material.

Significance – An object or a place that has importance to society, a cultural group, or an archaeologist conducting an investigation.

Site – A place where people conducted activities such as a camp or village.

Site Characterization – Using different methods to determine the presence, absence,or nature of resources at a location, e.g., subsurface archaeological deposits.

Site Type – Identifying an archaeological site by assigning function or a relative age such as a Clovis hunting camp; a site identified by the type of artifacts representative of an activity.

Stabilization – Usually methods for keeping the soil from moving around by covering or revegetating.

Stewardship – Taking the responsibility to care for the land or some other thing of importance.

Stratigraphy – The study of the strata of a site; the layers of deposits at a site hold information about the land and the people who lived there.

Stylistic Analysis – Scientific studies to determine cultural preferences of the forms of artifacts used by a group of people. Factors affecting artifact style include function, raw material, and technological constraints.

Subsurface Testing – Using a variety of methods to determine archaeological remains below the ground surface.

Survey – The systematic search for cultural materials across the landscape. Cultural materials are then recorded on a map. Surveys can be conducted on foot or with remote sensing tools that allow below-ground mapping.

Testing – Subsurface locations are often investigated using shovel probes or augers to identify archaeological materials in early stages of a project.

Traditional Cultural Property (TCP) – A location that has traditional importance or significance to a group of people. There may be no physical evidence of use.

Transect – A term used in cultural resources surveys., it is the line followed by a survey crew member during a survey or reconnaissance; strategies involve spacing between transects at a standard distance (10–15 m apart) so that as much of the project area as possible is observed.

Undisturbed – Usually refers to a subsurface archaeological context that has not been impacted by historic or modern land use, i.e., substantial earth moving, excavation, and building.

Unit – A defined space usually within a grid system, often used by archaeologists to designate excavation areas.

Use-wear – Recognizable on an artifact created as a result of use.

Utilized flake – A lithic flake with evidence of use.

Village Site – A permanent or semi-permanent settlement with structures and a diverse artifact assemblage.

References

Advisory Council on Historic Preservation. 2009. "Working with Section 106." *www.achp.gov/work106.html,* accessed 5 June 2009.

Boulding, J. Russell, and Jon S. Ginn. 2004. *Practical Handbook of Soil, Vadose Zone, and Ground-Water Contamination: Assessment, Prevention, and Remediation.* Second edition. Lewis Publishers, Boca Raton, FL.

Burke, Heather, Claire Smith, Dorothy Lippert, Joe Watkines, and Larry Zimmerman. 2008. *Kennewick Man: Perspectives on the Ancient One.* Left Coast Press, Walnut Creek, CA.

Burney, Michael S., and Wendy R. Sampson. 2004. "The Village and Cemetery of Tse-Whit-Zen: Lower Elwha Klalla, Occupation of Ediz Hook and the Port Angeles Harbor, Clallam County, Washington," Barbara L. Scott, editor (No. 4, November 9). Prepared for the Lower Elwha Klallam Tribe, Port Angeles, Washington. On file with the same and Burney & Associates, Taos, NM.

Burney, Michael S., and Angela M. Scarlata. 2003. "The Ihanktonwan's Sacred Place Called White Swan in the White Swan State Recreation Area, and East White Swan in the North Point Recreation Area, Lake Francis Case, Yankton Sioux Reservation, Charles Mix County, South Dakota." Prepared for the Yankton Sioux Tribe, Marty, South Dakota. On file, South Dakota State Archaeological Research Center, Rapid City, SD.

Burney, Michael S., and Jeff Van Pelt (editors). 2003. "Its About Time, It's About Them, It's About Us." *Journal of Northwest Anthropology*, Memoir No. 6. Moscow, ID.

City of Sacramento. 2003. "Native American Artifacts Found while Excavating." *City Hall Refurbishment and Expansion Project Newsletter* 1(4), July–August. City of Sacramento, Department of Public Works, Sacramento, CA.

Engber, Daniel. 2005. "How Do You Move a Cemetery?" *Slate Magazine,* 1 September 2005. Washington Post, *www.slate.com.*

Galbraith, Kate. 2007. "New Homes Confront Old Burial Grounds." *New York Times,* 17 June 2007.

ICOMOS (International Council on Monuments and Sites). 1990. *Charter for the Protection and Management of the Archaeological Heritage*. Prepared by the International Committee for the Management of Archaeological Heritage (ICAHM) and approved by the 9th General Assembly in Lausanne in 1990. *www.icomos.org*.

———. 2008. International Council on Monuments and Sites. *www.icomos.org*, accessed 15 December 2008.

JLARC (Joint Legislative Audit and Review Committee). 2006. *Review of Port Angeles Graving Dock Project*. Report 06–8. Transportation Performance Audit Board. State of Washington. Olympia. *www.leg.wa.gov/reports/06–8.pdf*, accessed 11 July 2009.

Killion, Thomas W. 2008. *Opening Archaeology: Repatriation's Impact on Contemporary Research and Practice*. SAR Press, Santa Fe, NM.

King, Thomas F. 2009. *Our Unprotected Heritage: Whitewashing the Destruction of Our National and Cultural Environment*. Left Coast Press, Walnut Creek, CA.

———. 2008. *Cultural Resource Laws & Practice*. Third edition. AltaMira Press, New York.

———. 2007. *Saving Places That Matter: A Citizen's Guide to the National Historic Preservation Act*. Left Coast Press, Walnut Creek, CA.

———. 2005. *Doing Archaeology: A Cultural Resource Management Perspective*. Left Coast Press, Walnut Creek, CA.

———. 2003. *Places That Count: Traditional Cultural Properties in Cultural Resource Management*. AltaMira Press, Walnut Creek, CA.

———. 2001. *Federal Planning and Historic Places: The Section 106 Process*. AltaMira Press, Walnut Creek, CA.

KCRA. 2003. "Sacramento City Hall Delay Costs $4 Million: Hundreds of Indian Artifacts Found On Construction Site." KCRA.com On-line News, *www.kcra.com/news/2563340/detail.html*, 17 October 2003. Article accessed 6 July 2008.

McGimsey, Charles R. III, and Hester A. Davis (editors). 1977. *The Management of Archeological Resources: The Airlie House Report*. Special Publication of the Society for American Archaeology, Washington, DC.

McGimsey, Chip, Leslie E. Eisenberg, and John F. Doershuk. 2003. Register of Professional Archaeologists Standards Board Decision. *www.rpanet.org*. Report accessed 6 June 2008.

Melmer, David. 2003. "Judge Rules against Yankton Sioux Tribe: Sacred Ground to Become Campground." *Indian Country Today*, 30 April 2003, p. B1.

Neumann, Thomas W., and Robert M. Sanford. 2001. *Cultural Resources Archaeology*. AltaMira Press, Walnut Creek, CA.

Nielson, D. M. 2006. *A Practical Handbook of Environmental Site Characterization and Groundwater Monitoring*. CRC Press, Boca Raton, FL.

Sacramento Bee. 2003a. "Hearing to Halt Construction of New City Hall is Delayed." 30 May 2003. Sacramento, CA.

———. 2003b. "City Hall Project Challenged: Work Has Disturbed an Indian Burial Site." 29 May 2002. Sacramento, CA.

———. 2003c. "Indian, City Reach Deal on Remains." 14 June 2003. Sacramento, CA.

Sacramento County. 2006. *American River Parkway Plan*. Sacramento County Department of Planning and Community Development, Sacramento, CA.

Sprague, Roderick. 2005. *Burial Terminology: A Guide for Researchers*. AltaMira Press, Walnut Creek, CA.

Stapp, Darby C., and Michael S. Burney. 2002. *Tribal Cultural Resource Management: The Full Circle to Stewardship*. AltaMira Press, Walnut Creek, CA.

State of Washington. 2007. Fact Sheet for NPDES Permit 002264–1. Blaine Wastewater Treatment Plant. Washington State Department of Ecology, Olympia.

UNESCO 2008. Culture. United Nations Educational, Scientific, and Cultural Organization. *http://portal.unesco.org/culture/en/ev.php-url_id=34603&url_do=do_topic&url_section =201.html,* accessed 15 December 2008.

Wheaton, T. "CRM at the World Archaeological Conference Number 4." *Society for American Archaeology Bulletin* 16(5). Society for American Archaeology, Washington, DC.

White, Gregory G., and Thomas F. King. 2007. *The Archaeological Survey Manual*. Left Coast Press, Walnut Creek, CA.

White, Tim D., and Pieter A. Folkens. 2005. *Human Bone Manual*. Academic Press, New York.

World Archaeological Congress. 2008. *www.worldarchaeologicalcongress.org/site/home.php*, accessed 16 November 2008.

World Bank. 1994. *Environmental Assessment Sourcebook*. Cultural Heritage in Environmental Assessment. Update Number 8. September. *http://go.worldbank.org/CR3COCT8PO*, accessed 16 November 2008.

———. 1999. *Environmental Assessment Sourcebook*. Chapter 3. Social and Cultural Issues in Environmental Review. Permanent URL for this page: *http://go.worldbank.org/CR3COCT8PO*, accessed 16 November 2008.

WSOT (Washington State Department of Transportation). 2006. *The Hood Canal Bridge Rehabilitation Project and Graving Dock Program. A Report to the Governor and Legislature of the State of Washington*. May 16, 2006. Washington State Transportation Committee, Olympia, WA.

INDEX

About the Authors

Darby C. Stapp (B.A. University of Denver; M.A. University of Idaho; Ph.D. University of Pennsylvania) and **Julia G. Longenecker** (B.A., University of Wyoming; M.A., University of Idaho) began their archaeological careers during the 1970s in the Rocky Mountain region of North America. Since then, they have worked in Gujarat, India, and the U.S. states of Colorado, Georgia, Idaho, Mississippi, Oregon, Pennsylvania, Texas, Washington, and Wyoming. They have worked for a variety of institutions, including the Smithsonian National Museum of Natural History; the University Museum, University of Pennsylvania; and universities, state and federal agencies, and American Indian tribes across the American West. Their work over the last twenty years has encompassed all aspects of construction projects, numbering in the thousands, to ensure protection of archaeological and other heritage resources. They have been directly involved in several high-profile archaeological disasters, including the Kennewick Man case, the Blaine Wastewater Treatment Plant Expansion project, and the Manhattan Project landfill project. Both are Registered Professional Archaeologists (*www.rpanet.org*). Additional information about the authors can be found at their website: *NorthwestAnthropology.com*.